"十四五"职业教育国家规划教材

Practical English for Energy Engineering Industry

涉外能源工程实用英语

刘清 主编

王国宏 李风山 副主编

化学工业出版社

·北京·

内容简介

《涉外能源工程实用英语》基于涉外能源工程作业现场以及典型涉外工作环节需要而编写。全书按照10个情境分别介绍了必备词汇、惯用单句、情景对话及模拟演练任务等工作过程的标准成套英语会话内容，涉及前期准备、工程进度安排、检查验收设备、安全吊装设备、资料图纸讨论、设备安装调试与试车拆检维修、现场操作培训等场景。本书具有较强的针对性、实用性和通用性，也符合当今国际通用施工惯例。书中还介绍了能源工程项目中必备的英语词汇及表达，作为语言交流的基础知识支撑。

本书既可作为从事涉外工程施工人员或生产操作人员的参考用书，也可作为职业院校能源类专业行业英语教材及企业相关人员技术培训用书。

图书在版编目（CIP）数据

涉外能源工程实用英语/刘清主编. —北京：化学工业出版社，2019.2（2023.8重印）
ISBN 978-7-122-33803-7

Ⅰ.①涉… Ⅱ.①刘… Ⅲ.①能源-英语 Ⅳ.①TK

中国版本图书馆CIP数据核字（2019）第011668号

责任编辑：旷英姿　王海燕　　　　　　　　装帧设计：王晓宇
责任校对：王素芹

出版发行：化学工业出版社（北京市东城区青年湖南街13号　邮政编码100011）
印　　装：涿州市般润文化传播有限公司
787mm×1092mm　1/16　印张12¾　字数202千字　2023年8月北京第1版第4次印刷

购书咨询：010-64518888　　　　　　　　　　售后服务：010-64518899
网　　址：http://www.cip.com.cn
凡购买本书，如有缺损质量问题，本社销售中心负责调换。

定　价：38.00元　　　　　　　　　　　　　　　　　　　版权所有　违者必究

编审人员名单

主　　编　刘　清
副 主 编　王国宏　李风山
编写人员　(按姓名笔画排序)
　　　　　丁永福　马　刚　王国宏　尹少荣　刘　清
　　　　　许焕强　李风山　李亚婷　辛艳萍　张旭升
　　　　　岳　峰　孟卫宁　姜　永

前言 Foreword

能源行业是我国经济发展的重要组成部分。近年来，为提升企业核心技术和装备水平，提高企业国际竞争力，能源行业加快转型升级速度，企业积极践行"一带一路"高质量发展建设，加强国际合作，互联互通，外籍工程技术人员来华进行技术合作、设备安装、维修养护等交流活动日益频繁。但是，目前精通岗位技能的一线技术人员英语实用会话能力普遍存在不足，在作业过程中经常由于语言交流困难导致工作不畅。鉴于上述实际问题，基于"培养"与"需求"精准对接的人才培养理念，由能源行业资深专家、一线技术人员、翻译人员、资深英语教师和专业课教师组建教材编写团队，通力合作，编写了《涉外能源工程实用英语》一书，书中的语言情境设计和相关案例都源于工作现场，力求实现职业英语与工作岗位实际需求的深度融合，方便学习者更好地学以致用。本书通俗易懂、简单易学、实用性强，这将对一线工程技术人员和能源类职业院校在校生学习适应涉外工作岗位要求的英语语言能力，提升岗位胜任能力和个人可持续发展起到极大的推动作用，为提高能源企业经济效益和国际化竞争力提供人才保障。

本书基于能源涉外工程作业现场，对施工过程中涉及的各个典型环节进行全方位的描述，包括：与外方人员初次见面介绍与问候、带领参观介绍企业、现场工作进度计划安排与讨论、与外方人员一起开箱检查验收进口设备、进口设备吊装搬运安全注意事项、设备吊装前与外方人员讨论技术图纸资料和说明书、与外方人员一起吊装进口设备、与外方人员一起安装调试试车拆检和维护、外方现场操作培训准备交接和合作庆祝等情景。这些情景构成基于工作过程的标准成套英语会话内容，具有很强的针对性、实用性和通用性，也符合当今国际通用施工惯例，易于学习者把语言知识运用到涉外能源项目工程交流中。书中还编写了自主学习知识储备内容，为学习者提供了在涉外能源工程项目中所涉及的应知应会的必备表达、词汇及术语，以此作为语言交流的基础知识支撑。这部分内容具有跨学科、跨领域的特点，解决学习者的语言交流知识面窄、英语词汇使用范围有限的问题。

本书第 4 次印刷融入党的二十大报告中职业教育"产教融合、科教融汇"的理念，致力于培养学生具有全球视野，通晓能源化工行业国际标准，精通行业技能和行业英语的高技能人才。

特别感谢能源行业龙头企业——国家能源集团宁夏煤业公司的领导及技术人员在本书编写过程中给予的大力支持与帮助。他们为本书提供了大量涉外能源项目工程英语现场实践案例和素材，使本书内容既源于岗位的实践，又适用于岗位语言能力需求，实现了人才"培养"与"需求"精准对接，是宁夏工业职业学院和国家能源集团宁夏煤业公司"产教融合、校企合作"成果的典范。

由于编者水平和经验有限，编写中如有不当与疏漏之处，请专家、读者批评指正。

编者

Scene One   Self-introduction & Greetings on Work Site
情景一   初次见面介绍及问候

Part Ⅰ   Background Information
        第一部分   情景背景知识 ·· 1

Part Ⅱ   Key Words and Expressions
        第二部分   关键词汇 ·· 2

Part Ⅲ   Typical Sentences
        第三部分   典型单句 ·· 3

Part Ⅳ   Situational Dialogues
        第四部分   实景开口说 ·· 4

Part Ⅴ   Exercises
        第五部分   练习活动 ·· 6

Scene Two   Plant Visit
情景二   参观工厂

Part Ⅰ   Background Information
        第一部分   情景背景知识 ··· 11

Part Ⅱ   Key Words and Expressions
        第二部分   关键词汇 ··· 20

Part Ⅲ   Typical Sentences
        第三部分   典型单句 ··· 20

Part Ⅳ   Situational Dialogues
        第四部分   实景开口说 ··· 21

Part Ⅴ   Exercises
        第五部分   练习活动 ··· 24

Scene Three   Project Schedule and Plan Discussion
情景三   工程进度及计划讨论

Part Ⅰ   Background Information
第一部分   情景背景知识 ·· 29

Part Ⅱ   Key Words and Expressions
第二部分   关键词汇 ·· 34

Part Ⅲ   Typical Sentences
第三部分   典型单句 ·· 35

Part Ⅳ   Situational Dialogues
第四部分   实景开口说 ··· 36

Part Ⅴ   Exercises
第五部分   练习活动 ·· 39

Scene Four   Unpacking and Inspecting Equipment with Foreigners
情景四   与外方人员一起开箱检查验收设备

Part Ⅰ   Background Information
第一部分   情景背景知识 ·· 45

Part Ⅱ   Key Words and Expressions
第二部分   关键词汇 ·· 55

Part Ⅲ   Typical Sentences
第三部分   典型单句 ·· 55

Part Ⅳ   Situational Dialogues
第四部分   实景开口说 ··· 56

Part Ⅴ   Exercises
第五部分   练习活动 ·· 58

Scene Five   Safety Notes on Hoisting and Handling Equipment
情景五   设备吊装搬运安全注意事项

Part Ⅰ   Background Information
第一部分   情景背景知识 ·· 64

Part Ⅱ	Key Words and Expressions
	第二部分　关键词汇 ·· 71
Part Ⅲ	Typical Sentences
	第三部分　典型单句 ·· 72
Part Ⅳ	Situational Dialogues
	第四部分　实景开口说 ·· 73
Part Ⅴ	Exercises
	第五部分　练习活动 ·· 75

Scene Six　Discussing the Technical Documents, Drawings and Instructions with Foreigners before Lifting
情景六　设备吊装前与外方人员讨论技术资料、图纸和说明书

Part Ⅰ	Background Information
	第一部分　情景背景知识 ·· 80
Part Ⅱ	Key Words and Expressions
	第二部分　关键词汇 ·· 86
Part Ⅲ	Typical Sentences
	第三部分　典型单句 ·· 87
Part Ⅳ	Situational Dialogues
	第四部分　实景开口说 ·· 88
Part Ⅴ	Exercises
	第五部分　练习活动 ·· 90

Scene Seven　Hoisting Imported Equipment with Foreigners
情景七　与外方人员一起吊装进口设备

Part Ⅰ	Background Information
	第一部分　情景背景知识 ·· 95
Part Ⅱ	Key Words and Expressions
	第二部分　关键词汇 ·· 100
Part Ⅲ	Typical Sentences
	第三部分　典型单句 ·· 101
Part Ⅳ	Situational Dialogues
	第四部分　实景开口说 ·· 102

| Part V | Exercises |
| | 第五部分　练习活动 ·· 105 |

Scene Eight　Installing, Testing, Starting-up, Disassembling and Maintaining with Foreigners
情景八　与外方人员安装、调试、开车、拆检和维护

Part Ⅰ	Background Information
	第一部分　情景背景知识 ·· 110
Part Ⅱ	Key Words and Expressions
	第二部分　关键词汇 ·· 120
Part Ⅲ	Typical Sentences
	第三部分　典型单句 ·· 120
Part Ⅳ	Situational Dialogues
	第四部分　实景开口说 ·· 121
Part Ⅴ	Exercises
	第五部分　练习活动 ·· 124

Scene Nine　On-site Operation Training
情景九　现场操作培训

Part Ⅰ	Background Information
	第一部分　情景背景知识 ·· 129
Part Ⅱ	Key Words and Expressions
	第二部分　关键词汇 ·· 132
Part Ⅲ	Typical Sentences
	第三部分　典型单句 ·· 133
Part Ⅳ	Situational Dialogues
	第四部分　实景开口说 ·· 133
Part Ⅴ	Exercises
	第五部分　练习活动 ·· 138

Scene Ten　Co-operation Celebration
情景十　合作庆祝

Part Ⅰ　Background Information
　　　　第一部分　情景背景知识 ·· 144
Part Ⅱ　Key Words and Expressions
　　　　第二部分　关键词汇 ·· 144
Part Ⅲ　Typical Sentences
　　　　第三部分　典型单句 ·· 146
Part Ⅳ　Situational Dialogues
　　　　第四部分　实景开口说 ··· 146
Part Ⅴ　Exercises
　　　　第五部分　练习活动 ·· 149

Common Expressions and Vocabularies of Typical Imported Equipment for Energy Engineering
能源工程典型进口设备常用表达及词汇

Part Ⅰ　Typical communication language for Daily work
　　　　第一部分　典型岗位日常工作交流用语 ································· 155
Part Ⅱ　Equipment, Parts & Related Fields' Terms
　　　　第二部分　设备、部件及相关学科领域专业术语 ···················· 160

References
参考文献

Scene One 情景一

Self-introduction & Greetings on Work Site
初次见面介绍及问候

Part I Background Information
第一部分 情景背景知识

Some engineers from a German equipment company will arrive at the site of the project according to the date listed in the following table. Liujie, as a receptionist, needs to arrange the hotel and work place, and ask about their requirements through phone calls. In the following work agenda, will those engineers build a working team with technician Cuipeng and other Chinese staff, and direct the equipment installation, commissioning and so on.

德国设备公司的工程师将于下表所列日期按计划到达项目现场，刘杰作为接待人员，需要做好接待准备，安排好相关入住酒店、工作地点，并通过电话联系外方人员了解其他相关需求。在接下来的工作中，外方人员将与现场中方工作人员及技术人员崔鹏等组建一个工作团队，现场指导设备安装、调试等工作。

No. 序号	Name 姓名	Company 公司名称	Nationality 国籍	On Site Arriving Time 到达现场时间	Hotel 入住酒店	Work Site 工作地点	Contact 联系方式	Receptionist 接待人员
1	Brown	MAN Diesel & Turbo Group 曼柴油机与透平集团公司	Germany 德国	2014.10.9	Ao Lisheng 奥立升酒店	Room 305, No. 2 Project Building 2号项目楼 305房间		Liu Jie 刘杰
2	Huebner	SIEMENS 西门子公司	Germany 德国	2015.4.14	Ao Lisheng 奥立升酒店	Room 305, No. 2 Project Building 2号项目楼 305房间		Liu Jie 刘杰
3	John	SIEMENS 西门子公司	Germany 德国	2015.6.30	Ao Lisheng 奥立升酒店	Room 305, No. 2 Project Building 2号项目楼 305房间		Li Qiang 李强
4	Bob	SIEMENS 西门子公司	Germany 德国	2015.8.4	Ao Lisheng 奥立升酒店	Room 305, No. 2 Project Building 2号项目楼 305房间		Li Qiang 李强

Part II Key Words and Expressions
第二部分 关键词汇

project manager	项目经理	site general representative	驻工地总代表
construction superintendent	工地主任	administrator	管理员
engineer	工程师	technician	技术员
economist	经济员	supervisor	检查员
foreman	工长、领班	worker	工人

Self-introduction & Greetings on Work Site
情景一 初次见面介绍及问候

续表

chemical engineering	化工工程	process	工艺
mechanical equipment	机械设备	electrical	电气
instrument	仪表	piping	管道
welding	焊接	furnace building	筑炉
corrosion prevention	防腐	thermal-insulation	保温
heating-ventilation	采暖通风	quality control	质量管理
electrician	电工	pipe layer	管工
welder	焊工	carpenter	木工
turner	车工	blacksmith	铁工
builder	建筑工人	erector	安装工人
riveter	铆工	rigger	起重工
concrete worker	混凝土工	engine-driver	工程司机
repair worker	维修工	direction	指导

Part Ⅲ Typical Sentences
第三部分 典型单句

1	Welcome to China.	欢迎你到中国来。
2	Welcome to our working site.	欢迎你到我们工地来。
3	Please allow me to introduce myself, my name is Liu Jie.	请允许我介绍自己,我是刘杰。
4	I hope we have a nice cooperation.	希望今后友好合作。
5	My technical specialty is mechanical equipment engineering.	我的技术专业是机械设备工程。
6	That sounds great.	听起来很棒。
7	Your advice will be highly appreciated.	请多指教。
8	Never mind, my pleasure.	没关系,很高兴为您服务。

Part IV Situational Dialogues
第四部分 实景开口说

Liu Jie: Hello, Mr. Brown. Welcome to our work site. I'm Liu Jie.
你好布朗先生。欢迎来到我们的工地。我是刘杰。

Brown: Nice to see you, Mr. Liu.
很高兴见到你,刘先生。

Liu Jie: Nice to see you, too.
很高兴见到你。

Brown: Well, I wish we have a good cooperation.
嗯,希望我们合作愉快。

Liu Jie: Definitely, let's work hard together.
当然,让我们共同努力吧。

Brown: Sure. This is my colleague, Mr. Huebner.
好的。这位是我的同事,胡博先生。

Huebner: Glad to meet you, Mr. Liu. I am a manager. (project manager/site general representative/construction superintendent/administrator/engineer/technician/economist/supervisor/foreman/worker) I come from Germany.
我是经理(项目经理、驻工地总代表、工地主任、管理员、工程师、技术员、经济员、检查员、工长、工人)。我来自德国。

Liu Jie: Glad to meet you, Mr. Huebner. I work in the Air-separation Engineering Department of ADS Coal-to-liquids Project (Market Department, Safety & Inspection Department). This is my colleague Cui Peng.
我在ADS间接煤液化项目空分工程部(市场部、安监部)工作。这是我的同事崔鹏。

Self-introduction & Greetings on Work Site
情景一 初次见面介绍及问候

Brown: Hi, how do you do!
　　　嗨，你好!

Cui Peng: Hi, how do you do!
　　　嗨，你好!

Brown: My technical specialty is mechanical equipment engineering (chemical engineering, process, electrical, instrument, piping, welding, furnace building, corrosion prevention, thermal-insulation, heating-ventilation, quality control). What is your specialty?
　　　我的技术专业是机械设备工程（化学工程、工艺、电气、仪表、管道、焊接、筑炉、防腐、保温、采暖通风、质量管理）。你的专业是什么？

Cui Peng: I am a mechanician (electrician, pipe layer, welder, carpenter, turner, blacksmith, builder, erector, riveter, rigger, concrete worker, engine-driver, repair worker).
　　　我是一个机械工（电工、管工、焊工、木工、车工、铁工、建筑工人、安装工人、铆工、起重工、混凝土工、工程司机、维修工）。

Huebner: It sounds great.
　　　听起来很棒。

Cui Peng: We really appreciate your suggestions and advices in the following days.
　　　我们非常感谢二位接下来几天给予我们意见和建议。

Brown: Never mind, my pleasure.
　　　没关系，很高兴为您服务。

Cui Peng: Anyway, thanks a lot. This road leads to our site. It's not far from here. Shall we go there right now?
　　　不管怎样，依然很感谢您。这条路通向工地，离这不远，我们现在过去可以吗？

Brown: Sure, let's go!
　　　当然可以，走吧！

Liu Jie: Let's go!
走吧！

Part V　Exercises
第五部分　练习活动

Task 1: Study the following words and match the words in left column with corresponding Chinese phrases or terminologies in the right column.

____ 1. market department	a. 项目经理
____ 2. work team	b. 工程师
____ 3. project manager	c. 工作团队
____ 4. contractor	d. 业主
____ 5. quality control	e. 外方人员
____ 6. engineer	f. 市场开发部
____ 7. foreign staff	g. 工长、领班
____ 8. owner	h. 质量管理
____ 9. foreman	i. 承包商
____ 10. equipment company	j. 设备公司

Task 2: Make sure you know how to spell and pronounce all the words above. Then write a sentence for each word. The sentence must show you understand the meaning of the word.

_____
_____
_____
_____
_____
_____

Self-introduction & Greetings on Work Site
情景一 初次见面介绍及问候

Task 3: Translate the following sentences into Chinese/English.

(1) Do you speak English?

(2) I can speak English only a little, do you understand me?

(3) I can read English only with the help of a dictionary.

(4) 我很抱歉,我的英语说得不好,但我能看懂英文资料。

(5) 你知道我们应该如何用英文表达这个意思吗?

(6) 请告诉我怎样拼读这个专业术语。

Task 4: Role Play

Follow the example of the conversation you have learned in this scene. Work together as a group to create an episode on this task theme. And then use role-play to practice your English communication skills on self-introduction & greetings on work site.

情景模拟演练任务书

班级	
任务主题	Self-introduction & Greetings on Work Site 初次见面介绍及问候
组名、成员	
演练内容	Use the expressions about this task with some communication skills(e.g. describing, introduction) to create a dialogue, an episode etc. 运用与本任务主题有关的语言表达和交流技能(如描述、介绍)创设情景会话或情景剧等。
演练目的	1. master the expressions about this task. 掌握本任务中惯用的英语表达。 2. master the communication skills in English. 掌握英语交流技能。 3. put them into application. 把所学付诸于实践应用。
小组成员承担的任务及角色	
任务完成情况及步骤	
小组任务完成综合体会(综合成员个人体会)	

情景模拟演练成果展示评分标准

考核内容	分值	得分	备注
一、设计思路表达是否切合任务主题	20		
二、内容是否积极、健康、向上	20		
三、在角色表演过程中是否投入、认真、代入感强	20		
四、编选的情景模拟要自然，符合基本的英语表达习惯，是否掌握正确运用交际用语的能力	20		
五、语音语调、手势表情、即兴反应	10		
六、小组成员协作与团队精神体现	10		

评价 \ 得分	小组最终得分	小组成员最终得分
小组互评（包括所有成员）		
教师点评		

Attachment: Episode

附：情景剧稿

Scene Two 情景二

Plant Visit
参观工厂

Part I Background Information
第一部分 情景背景知识

In the process of construction projects, especially large-scale projects, the construction has a direct effect on the economic development of a region or even a country. Therefore, during the implementation of the project, government departments, higher authorities, domestic and foreign counterparts, and contractors/suppliers at all levels always visit the site for the purpose of work guidance, communication, learning and other activities.

You must be warm and thoughtful to receive visitors. First of all, you can introduce the general situation of the project, process flow, and equipment to the visitor, so that they have a good impression on the project. During the reception process, you should use formal and appropriate languages.

在工程项目建设过程中，尤其是大型项目工程，它的建设直接影响（have a direct effect）一个地区乃至国家的经济发展。因此，在项目的实施过程中，政府机关部门、上级单位、国内外同行业单位及各级承包商/供应商等常常到现场参观，进行工作指导、交流、学习等活动。

接待来访者，一定要热情和周到（warm and thoughtful）。首先可以向客人介绍工程的概况、工艺流程、设备等情况，使其对项目有个好的印象

(a good impression)。接待过程中,言语上要注意做到正式得体(formal and appropriate)。

Ningdong Coal Chemical Base Profile   宁东煤化工基地简介

Ningdong Energy and Chemical Industry Base is located in Lingwu, Yinchuan of Ningxia Hui Autonomous Region. Its construction scope consists of planning area, 645 square kilometers, and long-range planning area, 2855 square kilometers. The former mainly includes Yuanyanghu, Lingwu, Hengcheng coal mines, Shigouyi coal field and a heavy chemical industry project area of 13.57 square kilometers. The construction headquarters and airview map of plant are shown in Figure 2.1 and Figure 2.2.

宁东能源化工基地位于宁夏回族自治区首府银川市灵武境内。基地规划建设范围分为远景规划区和规划区两部分。远景规划区面积约2855平方公

Figure 2.1   Project construction headquarters
图 2.1   项目建设指挥部

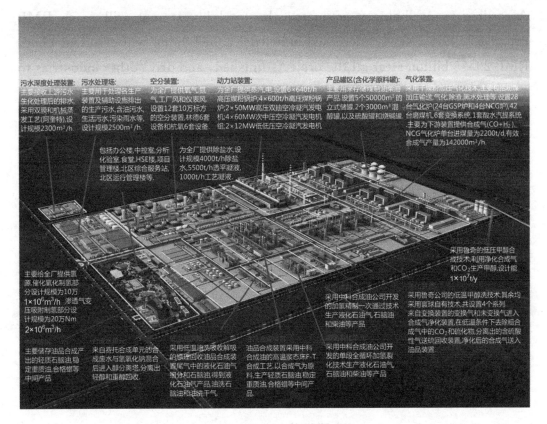

Figure 2.2 Airview map of plant

图 2.2 项目厂区鸟瞰图

里。规划区面积 645 平方公里，主要包括鸳鸯湖、灵武、横城三个矿区，石沟驿煤场及重化工项目区，其中重化工项目区规划面积 13.57 平方公里。基地项目建设指挥部和厂区鸟瞰图见图 2.1 和图 2.2。

The construction of the base is divided into two phases. The first phase is from 2003 to 2010 and the second from 2010 to 2020, including three industrial projects of coal, electric power and coal chemical industry, as well as infrastructure construction projects.

该基地建设分两期：一期为 2003 年到 2010 年，二期为 2010 年至 2020 年。包括煤、电力、煤化工三大产业项目和基础设施建设项目。

Overall objective：By 2020, Ningdong will achieve coal production capacity of 110 million tons, electric power installed capacity of 20 million kW, coal indirect liquefaction capacity of 10 million tons, coal-based dimethyl ether capacity of 2 million tons, and methanol capacity of 1.7 million tons.

According to preliminary estimate, the total investment will reach RMB 205.566 billion. Newly increased industrial added value will arrive at about RMB 29.76 billion after all the projects finished, and drive other industries to get production value of RMB 89.739 billion. By then, Ningdong Energy and Chemical Industry Base will become a national 10 million-kW thermal power base, coal chemical industry base and coal base with its pillar industries of coal, electric power and coal chemicals.

总体目标：到2020年，形成煤炭生产能力1.1亿吨，电力装机2000万千瓦以上，煤炭间接液化生产能力1000万吨，煤基二甲醚生产能力200万吨，甲醇生产能力170万吨。初步测算，基地总投资将达到2055.66亿元，全部项目建成后，将新增工业增加值约297.6亿元，并拉动其他行业形成产值897.39亿元。届时，宁东能源重化工基地将建设成为以煤炭、电力、煤化工三大产业为支撑，全国重要的千万千瓦级火电基地、煤化工基地和煤炭基地。

Ningdong Developing Superiority　　宁东发展优势

1. Policy Advantages　　政策优势

Ningdong base is listed as a national large-scale coal base, coal chemical industry base, circulated economic pilot garden, and one of 6 competitive industry bases in western China. The general planning of Ningdong four coal mines and the construction work of coal-to-liquid project are approved by China government. China Development Bank lists Ningdong as a national key supported project. With the further implement of "West Development" strategy, the state arranges the policy measures of supporting the west development, and Ningxia government makes great efforts to improve investment environment which provide the base construction with strong policy supports.

宁东基地被列入全国大型煤炭基地、煤化工产业基地、国家循环经济试点园区和西部地区六大优势产业基地之一，国家批准了宁东煤田4个矿区总体规划，批准在宁东建设煤制油项目工作。国家开发银行把宁东基地列为全国重点扶持项目。随着西部大开发战略的深入推进，国家实施重点支持西部

Scene Two Plant Visit
情景二 参观工厂

大开发的政策措施，以及宁夏政府全面改善投资环境的重大举措，为基地建设提供了强有力的政策支持。

2. Location Advantages 区位优势

Ningxia lies in the fringe area of north, northwest, and southwest of China, with limited geographical area and small population. It links resources and market. Located in the inevitable passage through which Gansu, Qinghai, Xinjiang and Inner Mongolia travel eastward, Ningdong Enengy and Chemical Industry Base is situated in the border area of Shanxi, Gansu, Ningxia, and Inner Mongolia, separated by the Yellow River with the capital city Yinchuan to the west, and sitting beside the developing Shanbei Energy and Heavy Chemical Industry Base to the east. It is helpful for mutual compensation of industries and sharing recourses.

宁夏地处华北、西北、西南地区的结合部，地域面积小，人口少，是资源和市场的连接带，又处在甘肃、青海、新疆、内蒙古等省区东出的必经通道上。宁东能源化工基地位于陕、甘、宁、蒙毗邻地区，西与宁夏回族自治区首府银川市隔黄河相望，东与开发中的陕北能源重化工基地毗邻，易形成产业互补，资源共享。

3. Coal Resource Advantages 煤炭资源优势

Ningdong coal field has proved reserves of over 27 billion tons, ranking the 6th in China. It is one of the 13 large-scale coal bases among the nation's priority of development. With abundant reserves, simple geologic structure, and good protections, main coal types of non-caking coal, coking coal and anthracite coal produced here are high-class chemical industry used coal and steam coal.

宁东煤田已探明储量270多亿吨，居全国第六位，是国家重点发展的13个大型煤炭基地之一，储量丰富，地质结构简单，保护较好。主要煤种为不黏结煤、炼焦煤和无烟煤，是优质化工用煤和动力用煤。

4. Transportation Advantages 交通运输优势

Extensive road traffic is an outstanding superiority of Ningdong Energy and Chemical Industry Base. Yinchuan-Qingdao expressway and the national road 307 traverse the base; Dagu railway linkes Baolan, Baozhong railway. When connected with Jingbao, Longhai railway, it radiates the whole country. Yinchuan-Taiyuan railway that is about to start construction will become another major outward channel. In addition, Yinchuan Hedong airport is just 30km away from the base center, and over 50 flights come and go every day, linking with important cities like Beijing, Shanghai, Guangzhou, Xi'an, Taiyuan, Ji'nan, Qingdao, and Lanzhou, etc.

四通八达的道路交通是宁东能源化工基地的一大突出优势，银川-青岛高速公路及307国道横贯基地；大古铁路连接包兰、宝中铁路与京包、陇海线连通可辐射全国，即将开工建设的银川-太原铁路又形成一条横穿基地的外运大通道；银川河东机场距基地中心区仅30公里，每日航班达50余次，通往北京、上海、广州、西安、太原、济南、青岛、兰州等重要城市。

5. Water Resource Advantages 水资源优势

Ningdong Energy and Chemical Base is located in the east bank of the Yellow River, about 35 km away from the Yellow River. The water supply construction began at the end of 2003, with the total water supply hitting 159.7 million cubic meters assuring water supply for the base. Thanks to the closeness to the Yellow River, the base can take water from the river instead of pumping groundwater for industry use. Through agricultural water-saving measures and the Yellow River water right transfer solutions, Ningdong never increases volume of water taking from the Yellow River which allocated to Ningxia by the government.

宁东能源化工基地位于黄河东畔，中心区距黄河仅35公里左右，2003年底开工建设的宁东供水工程，总供水量为15970万立方米，能为基地提供充足的水源保障。宁东地区紧靠黄河、取水方便，工业用水不抽用地下水，不增加国家给宁夏分配的黄河取水量，全部通过采取农业节水措施，以黄河水权转换方式解决。

6. Land Resource Advantages 土地资源优势

Ningdong Energy and Chemical Industry Base lies in the barren area, over 3,500 square kilometers of alkaline land and sandy wasteland with flat landform, open terrain, and adequate development site. Large-scale development will not occupy arable land. Its low cost of land development and no migrations provide vast land resources for the industrial construction.

宁东能源化工基地处于荒山丘陵地带,有3500多平方公里盐碱和沙荒地,地形平缓,地势开阔,有成片的发展用地,大规模开发建设不占用耕地、无移民搬迁、土地开发成本低,为工业建设提供了广阔的土地资源。

Figure 2.3 shows products exhibition of coal chemical project, and foreign engineers visiting plant is shown in Figure 2.4。

图 2.3 为煤化工项目产品陈列,图 2.4 为外方工程师参观工厂。

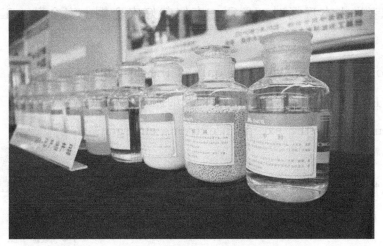

Figure 2.3 Products exhibition of coal chemical project
图 2.3 煤化工项目产品陈列

Figure 2.4 Foreign engineers visiting plant
图 2.4 外方工程师参观工厂

BASIC SAFETY REQUIREMENTS FOR THE VISITORS AND THE TEMPORARY WORKERS

参观人员、访客及临时工作人员的基本安全要求

(Workers and visitors must be trained by CNCPP HSSE prior to entering into the jobsite)

(参观人员、临时工作人员必须经神华宁煤项目相关安全培训后方可进入施工现场)

1. Hard hat, safety shoes, long sleeves shirt and the security badge must be worn at all times when visitors visit this site.

   当参观者进入本现场时，必须一直佩戴好安全帽、安全鞋、工作服及出入证。

2. No smoking on site (smoking only at the designated area).

   现场禁止抽烟（仅可在指定地点如吸烟棚等吸烟点抽烟）。

3. No Running on site, except in extreme emergency.

   除紧急情况外，不得在现场内奔跑。

4. Speed limit on site for all type vehicle: Below 15 km.

   车辆限速：不得超过每小时15公里。

5. Drivers or operators of cars and equipment are not permitted to get off from their driving cab.

   车辆驾驶员或设备操作工离开驾驶室/操作室，必须佩戴合格的个人防护用品。

6. Report every accident/incident/near-miss and/or injury to SNCP for investigation to prevent recurrence.

   向 SNCP 报告任何意外事故/未遂事故/潜在危险或伤害。

7. No camera, alcohol and unauthorized drugs are allowed on site.

   现场不得饮用酒或含酒精类饮料、不得服用违禁药品；未经许可不得

拍照。

8. Horseplay, fighting, gambling, possessions of firearms are not permitted on site.

   现场禁止嬉闹、打架斗殴、赌博、持有火器等。

9. Keep the workplace in good order. Rubbish and waste materials must be disposed after work in time.

   保持作业场所整洁有序：建筑垃圾及废料按要求及时清理。

10. Dangerous work (such as confined space entry, rigging/lifting and any other open flame work etc.) must get specific training of the procedure by SNCP safety and a written permit be obtained from SNCP HSSE is required.

    从事危险性工作（诸如受限空间作业、吊装作业、明火作业等），必须得到 SNCP 安全部的特别培训并取得书面的许可证后方可进行。

11. In case of emergency, evacuate to the safety area following the instruction of SNCP or signs on.

    如遇紧急情况，服从 SNCP 的指挥或遵守安全标志撤离到安全区域。

12. Worker (s) and/or visitor (s) must be escorted by associated person.

    工作人员及/或参观人员必须由熟悉现场的相关人员全程陪同。

13. No hardcopy nor electronic information of site will bring off-site without proper approval from SNCP.

    未经 SNCP 批准，任何拷贝文件及电子信息不得带出场地。

PRINT NAME 正楷姓名	COMPANY 公司
I have received a copy of your safety rules, and understand that I am required to observe these rules and any other safety instructions received while in this jobsite. 本人确认已收到贵现场的相关安全要求，本人理解并愿遵守这些安全规定及其他相关的安全指示	
SIGNATURE 签名　　A.Schnell MAN	DATE 日期

Part II Key Words and Expressions
第二部分　关键词汇

operator	操作员	production line	生产线
hectare	公顷	process	工艺
ASU(air separation unit)	空分装置	gasifier	气化炉
CO-shift unit	一氧化碳变换装置	rectisol unit	低温甲醇洗装置
SRU(sulfur recovery unit)	硫回收装置	F-T(fischer tropsch)unit	费托合成装置
oil processing unit	油品加工装置	tail-gas treatment unit	尾气处理装置
diesel	柴油	LPG(liquefied petroleum gas)	液化气
naphtha	石脑油	solid sulphur	固体硫黄

Part III Typical Sentences
第三部分　典型单句

1	Glad to see you.	很高兴见到你。
2	Do we need to put on the jackets?	需要穿夹克吗？
3	Please watch your step.	请注意脚下。

续表

4	How large is the plant?	工厂占地多大?
5	It covers a total area of almost 340 hectares.	厂区大约340公顷(1公顷为0.01平方千米)。
6	What are the major products of your plant?	工厂的主要产品有哪些?
7	Your company is so fantastic in developing coal chemical industry.	贵公司在发展煤化工方面真是棒极了。
8	Thanks for your praise!	谢谢你的夸奖!

Part Ⅳ Situational Dialogues
第四部分 实景开口说

Liu Jie: Welcome to our project, Mr. Brown and Mr. Huebner.
　　　　Brown先生、Huebner先生,欢迎来我们项目组。

Brown \ Huebner: Thank you.
　　　　谢谢。

Liu Jie: This is site engineer and operator, Mr. Chen Jun.
　　　　这是现场工程师也是操作员,陈俊先生。

Brown \ Huebner: Glad to see you, Mr. Chen.
　　　　陈先生,很高兴见到你。

Chen Jun: Glad to meet you too, Mr. Brown and Mr. Huebner. Put on the helmet, please.
　　　　布朗先生、胡博纳先生,很高兴见到你们。请戴好安全帽。

Brown: Do we need to put on the jackets, too?
　　　　我们也需要穿夹克吗?

Liu Jie: You'd better, to protect your body. Now please watch your step.
　　　　建议你穿上,可以保护你。请注意脚下!

Brown: Thank you. Is the production line fully automated?

　　　　谢谢。生产线是全自动化吗？

Chen Jun: Well, not fully automated, around 90%.

　　　　　不完全是，自动化大概占90%。

Brown: How large is the plant?

　　　　工厂占地多大？

Chen Jun: It covers a total area of almost 340 hectares.

　　　　　厂区大约340公顷。

Liu Jie: This plant is located in Ningdong Energy and Chemical Industry Base.

　　　　工厂位于宁东能源化工基地。

Huebner: Sounds great, Ningdong Energy and Chemical Industry Base? Could you please share more information with us?

　　　　　听起来不错，宁东能源化工基地？能分享更多的信息吗？

Liu Jie: Sure, Ningdong Energy and Chemical Industry Base is located in Lingwu, Yinchuan of Ningxia Hui Autonomous Region. Its construction scope consists of planning area, 645 square kilometers, and long-range planning area, 2855 square kilometers. The former mainly includes Yuanyanghu, Lingwu, Hengcheng coal mines, Shigouyi coal field and a heavy chemical industry project area of 13.57 square kilometers.

　　　　当然可以，宁东能源化工基地位于宁夏回族自治区首府银川市灵武境内。基地规划建设范围分为远景规划区和规划区两部分。远景规划区面积约2855平方公里。规划区面积645平方公里，主要包括鸳鸯湖、灵武、横城三个矿区、石沟驿煤场及重化工项目区，其中重化工项目区规划面积13.57平方公里。

Huebner: I See. How many process units involved in your plant?

　　　　　明白了。这个工厂包含多少工艺装置？

Chen Jun: The process production units mainly include twelve sets of ASU unit, twenty-eight gasifiers, six sets of CO-shift unit, four sets of rectisol unit, three sets of SRU, eight sets of F-T unit, one set of oil processing unit and tail-gas treatment unit.

工艺生产装置主要包括 12 套空分装置、28 台气化炉、6 套一氧化碳变换装置、4 套低温甲醇洗装置、3 套硫回收装置、8 套费托合成装置、1 套油品加工装置及 1 套尾气处理装置。

Brown: What are the major products of your plant?

工厂的主要产品有哪些？

Chen Jun: Generally, it makes coal to liquids, including: diesel, LPG, naphtha, solid sulphur and other by-products. The total volume is up to 4 million tons per year.

大体来讲，是把煤液化制成油品，包括：柴油、液化气、石脑油、固体硫黄及其他副产品。总产量达到每年 400 万吨。

Brown: It's cool, so huge.

太棒了，如此巨大。

Liu Jie: This plant is one of the large-scale coal chemical plants in Ningdong Energy and Chemical Industry Base. This Base is Ningxia's "Project No. One". ADS company is the main force of building Ningdong Energy and Chemical Industry Base, and takes the responsibilities of achieving the economic prosperity of Ningxia.

这个工厂是宁东能源化工基地大型煤化工项目之一。宁东能源化工基地是宁夏的"一号工程"。ADS 公司是宁东能源化工基地建设的主力军，承担着宁夏经济繁荣发展的重任。

Brown: Your Company is so developed in coal chemical industry.

贵公司在发展煤化工方面真是太先进了。

Liu Jie: It is. Well, thank you so much for your visiting. If you have any questions or suggestions, please feel free to tell us.

确实是。嗯，非常感谢您的到访。如果您有任何问题或建议，请随时告知我们。

Brown: Of course we will. Thanks for your accompany.
当然会的。感谢你们的陪同。

Liu Jie: You're welcome!
不客气!

Part V Exercises
第五部分 练习活动

Task 1: Study the following words and match the words in left column with corresponding Chinese phrases or terminologies in the right column.

____ 1. production line	a. 计算机集成制造系统
____ 2. technique process	b. 气化炉
____ 3. manipulator	c. 空分装置
____ 4. gasifier	d. 尾气处理装置
____ 5. computer integrated manufacturing system, CIMS	e. 空压机
	f. 技术过程
____ 6. Air Separation Unit, ASU	g. 机器人操作器
____ 7. tail-gas treatment unit	h. 空冷塔
____ 8. coal field	i. 生产线
____ 9. Coal Chemical Base	j. 煤化工基地
____ 10. Air cooling tower	k. 煤田
____ 11. Main Air Compressor, MAC	

Task 2: Make sure you know how to spell and pronounce all the words above. Then write a sentence for each word. The sentence must show you understand the meaning of the word.

_____

_____

Task 3: Try your best to use the real work situation to list the basic safety requirements briefly for the visitors/workers entering into the work site.

_____
_____
_____
_____
_____

Task 4: Role Play.

Follow the example of the conversation above. Work together as a group to create an episode on this task theme. Then use role play to practice your English communication skills on plant visit.

情景模拟演练任务书

班级	
任务主题	Plant Visit 参观工厂
组名、成员	
演练内容	Use the expressions about this task with some communication skills (e. g. describing) to create a dialogue, an episode etc. 运用与本主题相关的语言表达(例如描述性语言)和交流技能创建情景对话或情景剧等。
演练目的	1. master the expressions about this task. 掌握有关此任务的表达方式。 2. master communication skills in English. 掌握英语交流技能。 3. put them into application. 把所学付诸于实践应用。
小组成员承担的任务及角色	
任务完成情况及步骤	
小组任务完成综合体会(综合成员个人体会)	

情景模拟演练成果展示评分标准

考核内容	分值	得分	备注
一、设计思路表达是否切合任务主题	20		
二、内容是否积极、健康、向上	20		
三、在角色表演过程中是否投入、认真、代入感强	20		
四、编选的情景模拟要自然，符合基本的英语表达习惯，是否掌握正确运用交际用语的能力	20		
五、语音语调、手势表情、即兴反应	10		
六、小组成员协作与团队精神体现	10		

评价＼得分	小组最终得分	小组成员最终得分
小组互评（包括所有成员）		
教师点评		

Attachment: Episode

附：情景剧稿

Scene Three 情景三

Project Schedule and Plan Discussion
工程进度及计划讨论

Part I Background Information
第一部分 情景背景知识

The owners and the constructors often hold some regular meetings (Figure 3.1) weekly or monthly. During the meetings, they make schedules for the following works, discuss various current matters, such as the project progress, quality, environmental protection, site security and safety precautions, etc.

在工程的实施过程中，通常业主与承包商定期或不定期地召开会议，通常为周例会或月例会（如图3.1），安排下一阶段的工作，并就存在的问题协商解决，例如进度、质量、环境保护、现场安全及安全措施等。

1. Project Schedule Tips　工程进度安排注意事项

For all the geometrical dimensions such as centerline position, coordinate position, levelness, plumbness, elevation and relative distance, strict construction requirements and acceptance codes shall be applied. Necessary protective measures shall be taken for equipment unloading, storage, handling, lifting and finished product protection, etc. in the process of construction. Temporary transportation road, water source, power supply, lighting, firefighting facilities, major materials and construction

Figure 3.1 Project schedule seminar
图 3.1 项目计划讨论会

equipment, and manpower shall be sufficiently prepared and arranged before equipment installation.

设备安装施工中对于设备安装的中心线、坐标位置、水平度、垂直度、标高和相对距离等各种几何尺寸，应采用严格的施工要求和验收规范；施工过程中设备的卸车、存放、运输、吊装乃至产品保护等必须采取必要的保护措施。在设备安装前对临时建筑运输道、水源、电源、照明、消防设施、主要材料和机具及劳动力应有充分准备和安排。

2. Imported Equipment Installation Construction Procedure 进口设备安装施工步骤

The installation and construction Procedure of imported equipment is as follows.

进口设备的安装和施工步骤如下所示。

Scene Three Project Schedule and Plan Discussion
情景三 工程进度及计划讨论

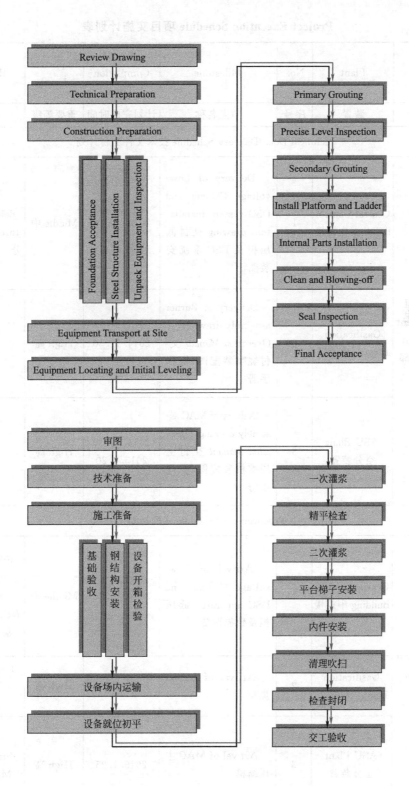

Project Execution Schedule 项目实施计划表

Department 部门	Plant 装置	No. 序号	Milestone 节点名称	Planned Completion Time 计划完成时间	Important Level 重要等级	Remark 备注
Technical Doc. Delivery Schedule 技术文件交付计划							
Technical Management Department. 技术管理部	Central Control Building 中控楼	1	Delivery of Low-voltage Cabinet and DSC system installation drawing 交付低压柜及 DSC 系统安装图	2014.10.16	Middle 中	To be delivered at three batches 分 3 批次交付
	Gasification Plant 气化装置	2	Delivery of Burner assembly drawing, *Operation Manual* 交付烧嘴装配图、操作手册	2014.11.30	High 高	
	ASU Plant 空分装置	3	Delivery of MAC assembly drawing, *Operation Manual* 交付主压缩机装配图、操作手册	2014.12.30 2015.8.20	High 高	
Procurement Schedule 采购计划							
Procurement Department. 采购部	Central Control Building 中控楼	1	Arrival of Low-voltage Cabinet and DSC apparatus 低压柜及数控设备	2014.10.30	Middle 中	Fifty 50 件 (including auxiliary facilities 包括辅助设施)
	Gasification Plant 气化装置	2	Arrival of burner 烧嘴	2015.12.15	High 高	Twenty sets of burner 20 套
	ASU Plant 空分装置	3	Arrival of MAC 主压缩机	2015.1.25	High 高	Six sets of MAC 6 套

Scene Three Project Schedule and Plan Discussion
情景三 工程进度及计划讨论

续表

Department 部门	Plant 装置	No. 序号	Milestone 节点名称	Planned Completion Time 计划完成时间	Important Level 重要等级	Remark 备注
Construction Schedule 施工计划							
System Engineering Department. 系统工程项目部	Central Control Building 中控楼	1	Completion of Low-voltage Cabinet installation 低压柜安装完成	2015.1.1	Middle 中	
		2	Completion of DCS apparatus installation DSC设备安装完成	2015.10.10	Middle 中	
		3	Ready for plant testing 达到全厂系统测试条件	2015.11.10	Middle 中	Depending on the completion of plant telecom system 取决于全厂电信系统完工
Construction Department. 施工部	Gasification Plant 气化装置	4	Completion of burner hoisting 烧嘴吊装完成	2015.6.30	High 高	Every two burners as one group (main, aided) 每2个一组（主、副）
		5	Completion of burner testing 烧嘴调试完成	2015.10.26	High 高	To conduct the burner testing frame by frame 以框架为单位进行调试
		6	Ready for burner ignition 达到点火条件	2015.1.18	High 高	

Department	Plant	No.	Milestone	Planned Completion Time	Important Level	Remark
部门	装置	序号	节点名称	计划完成时间	重要等级	备注
Construction Schedule 施工计划							
Construction Department. 施工部	ASU Plant 空分装置	7	Completion of MAC hoisting 主压缩机吊装完成	2015.6.26	High 高	
		8	Completion of MAC testing 主压缩机调试完成	2015.10.26	High 高	
		9	Ready for MAC operation 主压缩机达到投运条件	2015.12.18	High 高	Depending on the completion of ASU plant system engineering 取决于空分装置系统工程完工
OSBL Utilities Department. 厂外公用工程项目部	Logistics Center 物流中心	10	Ready for use 达到投运条件	2015.12.30	Low 低	
	Railway 铁路	11	Ready for use 达到投运条件	2015.12.30	Middle 中	

Part Ⅱ Key Words and Expressions
第二部分 关键词汇

schedule	计划	review	审查
account	报告、理由	feedback	反馈

Scene Three Project Schedule and Plan Discussion
情景三 工程进度及计划讨论

续表

contractor	承包商	procurement	采购
grace	宽限	overall schedule chart	总进度表
construction time schedule	建设进度表	claim	索赔
estimating	估算	trouble shooting	问题消除
expediting	加快	feasible	可行
daily report	日报（每日报告）	back-up	备用

Part Ⅲ Typical Sentences
第三部分 典型单句

1	Shall we start our meeting?	可以开会了吗？
2	Generally speaking, every activity has been going smoothly.	大体来说，各项工作进展都比较顺利。
3	The supplier of MAC requested 4 weeks grace for delivering the machine in its weekly report.	主压缩机供应商在周报中提出4周的交货宽限期。
4	We should work according to the overall schedule chart.	我们应该按照工程项目的总进度表工作。
5	If so, the late delivery will cause big cost during the period of waiting.	如果延迟，将在等待期间导致更多成本发生。
6	Our major planning items contain estimating of cost and construction schedule.	我们主要的计划工作项目包括费用预算和施工进度。
7	We shall have a progress collecting meeting next month.	下个月我们应召开进展信息收集会议。
8	It's good for expediting the progress of the project.	这有利于加快项目进展。
9	Six sets of MAC must be arrived at site before Jan. 25th, 2015.	6套主压缩机必须在2015年1月25日前到达现场。
10	Totally agree with you.	完全同意。

续表

11	We are going to start this tomorrow.	我们准备明天开始这项工作。
12	We have to consider another back-up plan, in case of the late delivery happening.	万一发生延迟交付,我们也必须考虑另一个备用方案。
13	What is your suggestion?	你的建议呢?
14	We should push the supplier forward, and give them more pressure.	我们应该催促供应商,给他们更大压力。

Part Ⅳ Situational Dialogues
第四部分 实景开口说

Liu Jie: Shall we start our meeting?
可以开会了吗?

Brown: Yes, go ahead please.
好的,开始吧。

Liu Jie: OK. Gentlemen, the purpose of our meeting today is to review and discuss the progress of this project, mainly relating to the ASU plant schedule. First of all, Mr. Brown, would you like to explain what you have achieved for the past month.
大家好,现在会议开始。今天会议的目的是审查并讨论项目的进度情况,主要是与我们有关的空分装置进度计划。首先,请Brown先生介绍下上个月的进展情况。

Brown: Yes. Generally speaking, every activity has been going smoothly, but there are certain problems. For the details, Mr. Huebner will give us an account of the progress for all the sections of the works.
大体来说,各项工作进展都比较顺利,但也存在一些问题。请Huebner介绍下具体工作过程中的详细情况。

Scene Three Project Schedule and Plan Discussion
情景三 工程进度及计划讨论

Liu Jie: Please, Huebner.
Huebner，请讲。

Huebner: According to the feedback from Technical Management Department, the delivery of MAC assembly drawing is under control now, so it could be submitted by the contactor on time. But in procurement aspect, it is not so good.
根据技术管理部的反馈，主压缩机装配图纸的交付工作承包商可以按时提交。但在采购方面，进展不是很好。

Liu Jie: What's the problem?
什么问题？

Huebner: I got an information from the Procurement Department, that is due to tight production schedule, the supplier of MAC requested 4-week grace period for delivering the machine in their weekly report.
主压缩机供应商在周报中提出4周的交货宽限期，主要原因是排产紧张。这是我从采购部得到的信息。

Brown: What? 4 weeks!
什么？4周！

Liu Jie: Oh, no. We should work according to the overall schedule chart (the construction time schedule) of the project.
啊，不行。我们应该按照工程项目的总进度表（建设进度表）工作。

Brown: I agree. Otherwise, the late delivery will impact on the following works related to hoisting and testing.
同意。否则，延迟交货将影响与之相关的吊装和后续测试工作。

Liu Jie: That's right. If so, the late delivery will cause big cost during the period of waiting. For you guys, it has to add more working hours, which will bring additional service fee. On other hand, the construction contractor will claim for such delay, due to the waste of working time.
是的。如果延迟，将在等待期间导致更多成本发生。对于你们，不得不增加工时数，产生额外的服务费；另一方面，施工承包商将提出误工索赔。

Chen Jun: Our major planning items contain estimating of cost and construction schedule.
我们主要的计划工作项目包括费用预算和施工进度。

Liu Jie: We shall have a progress collecting (trouble shooting) meeting next month.
下个月我们应召开进展信息收集（问题消除）会议。

Brown: It's good for expediting the progress of the project. And we should invite all related department to attend this meeting in order to have a feasible solution.
这有利于加快项目进展。同时，我们应该邀请所有相关部门出席会议，以便找到可行的解决方案。

Liu Jie: Sure. We must focus on the key target date as stipulated in ADS CTL project execution schedule.
对。我们必须紧盯 ADS 公司煤制油项目执行计划所确定的关键时间节点。

Chen Jun: Six sets of MAC must be arrived at site before Jan. 25th, 2015.
6 套主压缩机必须在 2015 年 1 月 25 日前到达现场。

Brown: Totally agree with you. Retain the arriving time of MAC is very important to complete the work of hoisting and testing as per the schedule in June and October 2015.
完全同意。维持主压缩机的交付时间对于依照计划在 2015 年 6 月和 10 月完成吊装及测试工作至关重要。

Liu Jie: We shall discuss the schedule every month, in order to keep everything goes smoothly according to the plan.
为了确保各项工作都能按计划顺利进行，每个月我们都要讨论进度计划。

Brown: I think it is necessary to ask the MAC supplier to make a daily report on its progress.
我认为非常有必要让主压缩机供应商提交每日工作报告。

Scene Three Project Schedule and Plan Discussion
情景三 工程进度及计划讨论

Huebner: We are going to start this tomorrow (next week).
我们准备明天（下周）开始这项工作。

Liu Jie: We must take this work plan immediately. Meanwhile, we have to consider another back-up plan, in case of the late delivery happening.
我们必须立即采取这个工作计划。同时，万一发生延迟交付，我们也必须考虑另一个备用方案。

Chen Jun: What are your suggestions, Mr. Brown and Huebner?
你们俩的意见呢？

Brown: We should push the supplier forward, and give them more pressure. I don't want to change our schedule for late delivery.
我们应该加紧催促供应商，不想因为延误而改变我们的进度计划。

Liu Jie: Of course, that's the worst case. Let's arrange a face-to-face meeting with the MAC supplier next Monday to talk about our concern.
当然，改变进度计划是最坏的情况。下周一安排一个与主压缩机供应商的面谈会议，谈一下我方的担忧。

Chen Jun: OK, I will contact the representative of the supplier, and let you all know, once it is confirmed with the supplier.
好的，我来联系供应商的代表，安排好了通知你们。

Liu Jie: Fine, that's all for this meeting.
好，今天的会就到这。

Part V Exercises
第五部分　练习活动

Task 1: Study the following words and match the words in left column with corresponding Chinese phrases or terminologies in the right column.

____	1. schedule	a.	中控楼
____	2. back-up	b.	总进度表
____	3. vendor	c.	订货单
____	4. subcontractor	d.	分包商
____	5. Central Control Building	e.	运输
____	6. transportation	f.	制造厂
____	7. technical preparation	g.	技术准备
____	8. review	h.	计划
____	9. purchasing order	i.	备用
____	10. overall schedule chart	j.	审查

Task 2: Make sure you know how to spell and pronounce all the words above. Then write a sentence for each word. The sentence must show you understand the meaning of the word.

_____
_____
_____
_____
_____
_____

Task 3: Try your best to use the real work situation to list the installation and construction procedure of imported equipment.

_____
_____
_____
_____

Task 4: Role Play

Follow the example of the conversation above. Work together as a group to

Scene Three Project Schedule and Plan Discussion
情景三 工程进度及计划讨论

draft a working schedule and create an episode on unpacking and inspection of the equipment, safety training, equipment hoisting, testing, operation training etc.. Then use role play to practice your English skills, so as to promote the competency at the work place.

情景模拟演练任务书

班级	
任务主题	Project Schedule Discussion 工程进度计划讨论
组名、成员	
演练内容	Use the expressions about this task with some communication skills(e. g. describing) to create a dialogue, an episode etc. 运用与本任务主题有关的语言表达和交流技能(如描述)创设情景会话或情景剧等。
演练目的	1. master the expressions about this task. 掌握有关此任务的表达方式。 2. master the communication skills in English. 掌握英语交流技能。 3. put them into application. 把所学付诸于实践应用。
小组成员承担的任务及角色	
任务完成情况及步骤	
小组任务完成综合体会(综合成员个人体会)	

情景模拟演练成果展示评分标准

考核内容	分值	得分	备注
一、设计思路表达是否切合任务主题	20		
二、内容是否积极、健康、向上	20		
三、在角色表演过程中是否投入、认真、代入感强	20		
四、编选的情景模拟要自然，符合基本的英语表达习惯，是否掌握正确运用交际用语的能力	20		
五、语音语调、手势表情、即兴反应	10		
六、小组成员协作与团队精神体现	10		

评价 \ 得分	小组最终得分	小组成员最终得分
小组互评（包括所有成员）		
教师点评		

Attachment: Episode

附：情景剧稿

Scene Four 情景四

Unpacking and Inspecting Equipment with Foreigners
与外方人员一起开箱检查验收设备

Part Ⅰ Background Information
第一部分 情景背景知识

Unpacking and inspecting of equipment shall be performed at the presence of the owner and supervising engineer. Inspection shall be carried out according to the packing list as follows: name of the equipment, type and model, specifications and quantity should be correct, overall dimensions of the equipment and nozzle orientation shall be in compliance with the drawing. Appearance inspection shall indicate no damage. After unpacking and inspecting, the "Record of Test for Equipment Unpacking" shall be filled out and signed by a related inspector immediately, and custody together with the attached documents and information.

设备的开箱检验必须有业主和监理工程师在场共同验收。根据装箱清单进行下列项目的检查：设备名称、型号、规格、数量应正确，设备总尺寸和管口方位与图纸相符。外观检查应无损伤，设备开箱检验完毕，应及时填写"设备开箱检验记录"，并由有关检验人员签字，随同附属资料一起保管。

1. Check upon arrival and unpacking 到货及开箱检查

(1) Check upon arrival 到货检查

The machine and the package must be inspected immediately upon arrival (showing in Figure 4.1~Figure 4.3). Any transport damage must be photographed and reported immediately, i.e. within less than one (1) week after arrival, if the transport insurance is to be claimed. It is, therefore, important that evidence of careless handling is checked and reported immediately to the transport company and the supplier.

必须在到货后立即检查机器和包装（如图 4.1~图 4.3）。任何运输导致

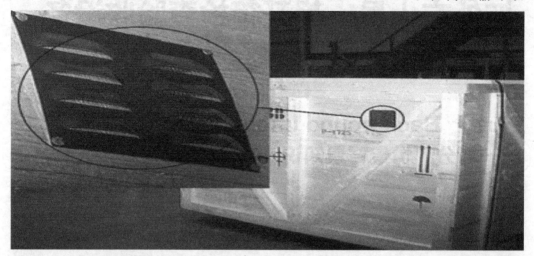

Figure 4.1　Package appearance inspection
图 4.1　包装外观检查

Scene Four Unpacking and Inspecting Equipment with Foreigners
情景四 与外方人员一起开箱检查验收设备

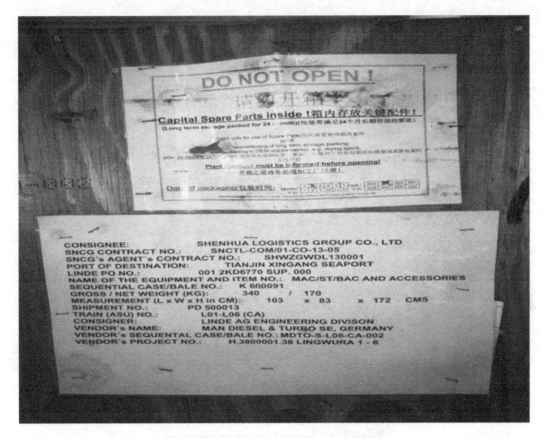

Figure 4.2 Imported equipment packing label
图 4.2 进口设备包装箱标签

的损坏应在收货后一周内拍照和上报。如果有运输保险，则在到达一周内可要求索赔。因此，重要的是应立即检查失误操作的证据，并向运输公司和供应商报告。

A machine, which is not to be installed immediately upon arrival, must not be left without supervision or without protective precautions.

如果在机器到货时不打算立即安装，必须设置监控或防护措施。

(2) Check upon unpacking 开箱检查

The machine should be placed on a flat and vibration-free surface, and make sure it won't affect the handling of any other goods.

机器的放置不得妨碍任何其他物品的处理并且应该放置在平坦和无振动的表面之上。

After the package has been removed, check that the machine is not dam-

Figure 4.3 Imported equipment unpacking
图 4.3 进口设备开箱

aged and that all accessories are included. Tick off the accessories on the packing list which is enclosed. If there is any suspected damage or if accessories are missing, take photographs and report this immediately to the supplier.

在包装拆除后，检查机器确保无损坏和所有的配件都包括在内。勾选附在包装清单上的配件。如果有任何疑似损坏或者配件丢失，请拍照并且立即向供应商报告。

2. Storage 储存

Short term storage (less than 2 months) 短期储存（不超过2个月）

The machine should be stored in a proper warehouse with a controllable environment. A good warehouse or storage place Should meet the fllowing requirements:

Scene Four Unpacking and Inspecting Equipment with Foreigners
情景四 与外方人员一起开箱检查验收设备

该机器应该存储在一个合适且可控的仓库环境中。一个合适的仓库或者存放地点需满足以下要求：

◆ A stable temperature, preferably in the range from 10℃ (50℉) to 50℃ (120℉). If the anti-condensation heaters are energized, and the surrounding air is above 50℃ (120℉), it must be confirmed that the machine is not overheated.

稳定的温度范围，最好在 10℃ (50℉) 和 50℃ (120℉) 之间。如果抗冷凝加热器通电以及环境温度在 50℃ (120℉) 以上，必须确认机器不会过热。

◆ Low relative air humidity, preferably below 75%. The temperature of the machine should be kept above the dew point, as to prevent moisture from condensing inside the machine. If the machine is equipped with anti-condensation heaters, they should be energized. The operation of the anti-condensation heaters must be verified periodically. If the machine is not equipped with anti-condensation heaters, an alternative method of heating the machine and preventing moisture from condensing in the machine must be used.

低相对空气湿度，最好低于 75%。机器的温度应该保持在水的露点以上，以便防止水分在机器内部冷凝。如果该机器配备有抗冷凝加热器，则应该对这些抗冷凝加热器通电。必须定期核查这些抗冷凝加热器的运行情况。如果该机器没有配备抗冷凝加热器，必须使用另一种方法来加热机器，防止水分在机器内部冷凝。

◆ A stable support free from excessive vibrations and shocks. If vibrations are suspected to be too high, the machine should be isolated by placing suitable rubber blocks under the machine feet.

不受过度振动和冲击影响的稳定支架。如果怀疑振动太大，应该在机脚下方放置合适的橡胶块，以便隔离机器。

◆ Ventilated and clean air free from dust and corrosive gases.
空气是流通的、干净的，并且无灰尘和腐蚀性气体。

◆ Protection against harmful insects and vermin.
防止有害昆虫和害虫。

If the machine needs to be stored outdoors, the machine must never be left 'as is' in its transportation package. Instead, the machine must be:

如果需要将机器存放在室外,则不得使机器处于运输包装中。相反,机器必须:

◆ Taken out from its plastic wrap.

从塑料包裹膜中取出。

◆ Covered, as to completely prevent from the rain. The cover should allow ventilation of the machine.

盖严实,以完全杜绝雨水进入。机盖可以确保机器能够良好通风。

◆ Placed on at least 100 mm high rigid supports, as to make sure that no moisture can enter the machine from below.

放置在至少 100 毫米高的刚性支架上,以确保水分不能从下面进入机器。

◆ Provided with good ventilation. If the machine is left in its transportation package, large enough ventilation openings must be made in the package.

提供良好的通风效果。如果机器在运输包装中,必须在包装上设置足够大的通风孔。

◆ Protected from harmful insects and vermin.

防止有害昆虫和害虫。

Long term storage (more than 2 months) 长期储存(超过 2 月)

In addition to the measures described with short term storage, the following should be applied:

除了短期储存中描述的措施外,还应该采取以下措施。

◆ Check the condition of the painted surfaces every three months. If corrosion is observed, remove it and apply a coat of paint again.

每三个月检查一次油漆面的状况。如果发现腐蚀,立即去除,然后再涂上一层油漆。

◆ Check the condition of anti-corrosive coating on blank metal surfaces (e. g. shaft extensions) every three months. If any corrosion is observed, remove it with a fine emery cloth and perform the anti-corrosive treatment again.

每三个月检查一次空白金属表面(轴外伸部)上的防腐涂料的状况。一

Scene Four Unpacking and Inspecting Equipment with Foreigners
情景四 与外方人员一起开箱检查验收设备

且发现任何腐蚀，用细砂布去除，然后再进行防腐处理。

◆ Arrange small ventilation openings when the machine is stored in a wooden box. Prohibit water, insects and vermin from entering the box, see the table below.

当将机器放在木箱内时，必须设置小通风孔。禁止水、昆虫和害虫进入箱内。参见下面列表。

Parts List

item no.	part number	description	quantity
10	S050159	tube	1
20	S051184	front end cover	1
30	S050303	rear end cover	1
40	S050157	piston	1
50	U100089-81	gasket	1
60	S051819	ram	1
70	0081-00-023-R	spring	1
80	S051897	stroke rod	1
90	U100200-38	tie rod	4
100	0032-89-M30	nut	8
110	U100200-98	yoke	1
120	U100200-40	coupling	1
130	0037-40-RB08O4	reducing bush	1
140	0037-21-MSE08L14BSPT	male stud elbow	2
150	U100252-43	positioner connector	1
160	U100200-44	positioner arm	1
170	0032-01-010	nut-full hex	8
180	0031-20-004010	ski headed cap screw	1
190	U100252-75	anti-rotation/positioner rod	1
200	0031-28-016055	ski headed cap screw	4
210	0039-3B-G2B	silencer	1

续表

item no.	part number	description	quantity
220	0031-20-006012	ski headed cap screw	5
230	0035-00-006	plain washer	4
240	0031-20-008016	ski headed cap screw	4
250	0032-89-M8	Nyloc nut	2
260	S047709	column	4
270	S050320	flange	1
280	S050306	anti-rotation plate	1
290	S050305	coupling	1
300	S044689	spring	1
310	0032-01-016	nut-full hex	4
320	0033-10-008050	spring dowel pin	1
330	0031-18-010065	M10 stud	2
340	0035-00-010	plain washer	2
350	0031-10-008090	M8 stud	1
360	0083-62-008	ball knob	1
370	0031-20-008030	ski headed cap screw	4
380	0037-21-MSE08L14NPT	male stud elbow	1
390	0075-11S-PS2-03	positioner	1
400	U100252-75-1	anti-rotation rod	1
410	0074-1S-G12-3	filter-regulator	1
420	U100252-79	positioner mounting plate	1
430	0037-40-RB0804N	reducing bush	1
440	0037-20-MSC08L14NPT	male stud coupling	1
450	0031-28-016070	ski headed cap screw	4
460	S050330	tail rod	1
470	E-0095-0U-U100200	sealing set	1

Scene Four Unpacking and Inspecting Equipment with Foreigners
情景四 与外方人员一起开箱检查验收设备

设备清单

项目编号	零件编号	类型	数量
10	S050159	管	1
20	S051184	前端盖	1
30	S050303	后端盖	1
40	S050157	活塞	1
50	U100089-81	垫片	1
60	S051819	柱塞	1
70	0081-00-023-R	弹簧	1
80	S051897	行程杆	1
90	U100200-38	拉杆	4
100	0032-89-M30	螺母	8
110	U100200-98	装配轭	1
120	U100200-40	联轴器	1
130	0037-40-RB08O4	缩口轴衬	1
140	0037-21-MSE08L14BSPT	外螺桩弯头	2
150	U100252-43	定位器连接头	1
160	U100200-44	定位器臂	1
170	0032-01-010	全六角螺母	8
180	0031-20-004010	六角螺钉	1
190	U100252-75	防转/定位器杆	1
200	0031-28-016055	六角螺钉	4
210	0039-3B-G2B	消音器	1
220	0031-20-006012	六角螺钉	5
230	0035-00-006	平垫圈	4
240	0031-20-008016	六角螺钉	4

续表

项目编号	零件编号	类型	数量
250	0032-89-M8	耐落螺母	2
260	S047709	柱	4
270	S050320	法兰	1
280	S050306	防转板	1
290	S050305	联轴器	1
300	S044689	弹簧	1
310	0032-01-016	全六角螺母	4
320	0033-10-008050	弹簧定位销	1
330	0031-18-010065	M10 螺柱	2
340	0035-00-010	平垫圈	2
350	0031-10-008090	M8 螺柱	1
360	0083-62-008	球头把手	1
370	0031-20-008030	六角螺钉	4
380	0037-21-MSE08L14NPT	外螺桩弯头	1
390	0075-11S-PS2-03	定位器	1
400	U100252-75-1	防转轩	1
410	0074-1S-G12-3	过滤器调节器	1
420	U100252-79	定位器安装板	1
430	0037-40-RB0804N	缩口轴衬	1
440	0037-20-MSC08L14NPT	外螺桩轴套	1
450	0031-28-016070	六角螺钉	4
460	S050330	导向杆	1
470	E-0095-0U-U100200	密封装置	1

Scene Four Unpacking and Inspecting Equipment with Foreigners
情景四 与外方人员一起开箱检查验收设备

Part II Key Words and Expressions
第二部分 关键词汇

packing list	装箱单	shipping list	发货清单
accessories	附件	bill of materials	材料清单
bill of lading	提货单	item number	项目编号
case number	箱号	by railway	铁路运输
by air	空运	by road	公路运输
symbol	符号	rollers	滚杠
heave here	从此处提起	handle with care	小心处理
top	上部	bottom	底部
inflammable	易燃	fragile	易碎
do not cast	勿掷	upright	勿倒置
keep in a cold (dry) place	存于冷（干）处	to be protected from cold (heat)	使其免受冷（热）

Part III Typical Sentences
第三部分 典型单句

1	Shall we go to the warehouse to check the equipments?	我们去仓库看一下设备吧？
2	I believe that all the machines shipped have been pre-inspected.	我相信所有发货的机器都曾经过预检验。
3	Where is the packing list?	装箱单在哪里？

续表

4	What is the item number of this equipment?	这台设备的编号是多少？
5	What does this mark mean?	这个标志是什么意思？
6	All the equipments must be fitted with nameplates.	所有的设备都钉有铭牌。
7	It has been damaged here, let's take a picture.	此处已经损坏，我们来照个相。
8	We should record the damaged and the missed parts.	我们应该把损坏和缺件情况作个记录。
9	The packaging has to be improved.	这个包装必须改进。
10	Please sign your name on the check list.	请您在检验单上签个字。

Part IV Situational Dialogues
第四部分 实景开口说

Liu Jie: Hi, Mr. Brown. Shall we go to the warehouse to check the equipments?
布朗先生，我们去仓库看一下设备吧？

Brown: Sure, I believe that all the machines shipped have been pre-inspected.
好的，我相信所有发货的机器都经过预检验。

Liu Jie: I believe so.
那肯定的。

Huebner: Let's check the quantity of the parts (accessories) according to the packing list (shipping list).
让我们根据装箱单（发货清单）来核查零件（附件）数量。

Liu Jie: Where is the packing list (bill of lading, bill of materials)?
装箱单（提货单、材料清单）在哪里？

Huebner: Oh, it's here.

Scene Four Unpacking and Inspecting Equipment with Foreigners
情景四 与外方人员一起开箱检查验收设备

哦，在这儿。

Liu Jie: What is the item number (case number) of this equipment?

这台设备的编号（箱号）是多少？

Brown: S050803. How many cases are there in the package of this equipment?

S050803，这台设备分几箱包装？

Liu Jie: One case. The equipment is delivered here by sea (air, railway, road) from Federal Republic of Germany. Oh, I see there are some marks on the package. What does this mark (symbol) mean?

1箱。这台设备是通过海运（空运、铁路、公路）从联邦德国发货到此的。哦，我看到包装上还有一些标志。这个标志（符号）是什么意思？

Huebner: Yes, there are some marks, and we should pay attention to these. For example:

是的，是有一些标志，我们要注意这些标志。例如：

Use rollers 使用滚杠移	Heave here 从此处提起
Handle with care 小心处理	Top 上部
Bottom 底部	Inflammable 易燃
Fragile 易碎	Do not cast 勿掷
upright 勿倒置	Keep in a cold (dry) place 存于冷（干）处

To be protected from cold (heat) 使其免受冷（热）

Liu Jie: Yeah, I see.

哦，我懂了。

Brown: All the equipments must be fitted with nameplates.

所有的设备都钉有铭牌。

Liu Jie: We plan on finishing the inspection of the equipments before next Sunday.

我们计划在下个星期天以前完成设备的检查工作。

Huebner: This is the equipment in completed conditions, which is the equipment fabricated at plant site.

这是完工状态下的设备，即在工地现场组装的设备。

Liu Jie: It has been damaged (rusted) here. Let's take a picture (photograph).

此处已经损坏（锈蚀），我们来照个相。

Brown: We should record the damaged and the missed parts.

我们应该把损坏和缺件情况作个记录。

Liu Jie: Shall we check it again (once more)? The packaging has to be improved.

我们再检查一次（再查一遍）好吗？这个包装必须改进。

Brown: Yes, but we have the spare parts. We must pack the spare parts into another box.

是的，但是我们有备件，我们应该将备件另行装箱。

Huebner: When I go back my country I will give feedback to our company, please sign your name on the check list.

我回国后会将此建议向公司反馈，请你在检验单上签个字。

Liu Jie: Ok, sign here?

好的，签在这儿吗？

Huebner: Yes, Please! Thanks.

是的，请签这儿，谢谢！

Part V Exercises

第五部分 练习活动

Task 1: Study the following words and match the words in left column with corresponding Chinese phrases or terminologies in the right column.

Scene Four Unpacking and Inspecting Equipment with Foreigners
情景四 与外方人员一起开箱检查验收设备

____ 1. unpacking	a. 储存
____ 2. warehouse	b. 易碎
____ 3. packing list	c. 提货单
____ 4. observe	d. 材料清单
____ 5. nameplate	e. 装箱单
____ 6. sampling inspection	f. 观察
____ 7. bill of materials	g. 开箱
____ 8. bill of lading	h. 铭牌
____ 9. fragile	i. 取样测定
____ 10. storage	j. 仓库
____ 11. surface roughness	k. 规格
____ 12. specification	l. 表面粗糙度
____ 13. inspector	m. 设备外形尺寸
____ 14. record of test for equipment unpacking	n. 检验人员
____ 15. overall dimensions of the equipment	o. 设备开箱检验记录

Task 2: Make sure you know how to spell and pronounce all the words above. Then write a sentence for each word. The sentence must show you understand the meaning of the word.

_____
_____
_____
_____
_____
_____
_____
_____
_____
_____

Task 3: Try your best to use the real work situation to list the steps of unpacking and inspection.

_____
_____

Task 4: Role Play.

Follow the example of the conversation above. Work together as a group to create an episode on this task theme. Then use role play to practice your English communication skills on unpacking and inspecting of the equipment with foreigners.

Scene Four Unpacking and Inspecting Equipment with Foreigners
情景四 与外方人员一起开箱检查验收设备

情景模拟演练任务书

班级	
任务主题	Unpacking and inspecting the Equipment with Foreigners 与外方人员一起开箱检查验收设备
组名、成员	
演练内容	Use the expressions about this task with some communication skills (e. g. describing) to create a dialogue, an episode etc. 运用与本任务主题有关的语言表达和交流技能(如描述)创设情景会话或情景剧等。
演练目的	1. master the expressions about this task. 掌握有关此任务的表达方式。 2. master the communication skills in English. 掌握英语交流技能。 3. put them into application. 把所学付诸于实践应用。
小组成员承担的任务及角色	
任务完成情况及步骤	
小组任务完成综合体会(综合成员个人体会)	

情景模拟演练成果展示评分标准

考核内容	分值	得分	备注
一、设计思路表达是否切合任务主题	20		
二、内容是否积极、健康、向上	20		
三、在角色表演过程中是否投入、认真、代入感强	20		
四、编选的情景模拟要自然，符合基本的英语表达习惯，是否掌握正确运用交际用语的能力	20		
五、语音语调、手势表情、即兴反应	10		
六、小组成员协作与团队精神体现	10		
评价 \ 得分	小组最终得分	小组成员最终得分	
小组互评（包括所有成员）			
教师点评			

Scene Four Unpacking and Inspecting Equipment with Foreigners
情景四 与外方人员一起开箱检查验收设备

Attachment: Episode

附：情景剧稿

Scene Five 情景五

Safety Notes on Hoisting and Handling Equipment
设备吊装搬运安全注意事项

Part I Background Information
第一部分 情景背景知识

  Any operation must be studied in details, particularly when the piece to be handled is heavy or large, in order to make sure that an adequate vehicle is used, the necessary room for crane is available, and to prevent any damage, especially to people, that could result from the fall of the load. If necessary, specialists must make sure of the proper condition of the vehicle used, or the evidence of the proper conditions (according to country regulations), as technical visit certificates, must be available. Only trained, qualified and authorized persons are allowed operate lifting or handling vehicles. Some operations may require the permanent presence of a signalman. The common safety signs and personal safety equipment are as follows:

  任何操作都必须详细研究，尤其是需搬运的零件较重或较大时，以确保使用适当车辆，有吊车所需的空间，并防止货物跌落造成损坏，尤其是人员损伤。如有必要，专家必须确保所用车辆状况良好或必须有证据证明其状况良好（符合国家条例）。只有经过培训合格，且经过授权的人员才能操作起重或搬运车辆。某些操作可能需要信号员全程在场。常见安全标志及个人安全设备如下：

Scene Five Safety Notes on Hoisting and Handling Equipment
情景五 设备吊装搬运安全注意事项

Safety signs   安全标志

Sign(符号)	Meaning(含义)	Sign(符号)	Meaning(含义)
	Warning of a hazard zone 危险区域警告		Warning of hazard to the environment 环境危害警告
	Warning of hot surface 热表面警告		Fire, open ignition source and smoking prohibited 禁止明火
	Warning of obstacles on the floor 地面障碍物绊倒警告		Do not use as lifting point 不得用作起吊点
	Warning of automatic starting 自动启动警告		Use eye protection 使用护眼设备
	Warning of hand injury 伤手警告		Wear hearing protection 佩戴听力保护设备
	Warning of pressurized housings and cylinders 加压气缸和外壳警告		Read the Operating Instructions 阅读操作说明书
	Warning of risk of unconsciousness/suffocation from inhalation of nitrogen 因吸入氮气无意识/窒息风险警告		Caution 注意安全

续表

Sign(符号)	Meaning(含义)	Sign(符号)	Meaning(含义)
	Warning of flammable substances 易燃物警告		Warning of risk of entanglement 缠绕风险警告
	Warning of danger from suspended load 悬载重危险警告		smoking prohibited 禁止吸烟
	Warning of risk of fall 跌倒风险警告		Wear a safety helmet, eye protection and hearing protection 佩戴安全帽、护眼设备和听力保护设备
	Warning of health-hazardous substances 健康有害物质警告		Emergency stop button 急停按钮
	Warning of high voltage 高电压警告		Authorized persons only 仅授权人员
	Warning of hot or harmful media emission 热或有害介质排放警告		Wear a safety helmet 戴上安全帽
	Warning of gas cylinders 高压气瓶警告		Wear hearing protection 佩戴听力保护设备

Scene Five Safety Notes on Hoisting and Handling Equipment
情景五 设备吊装搬运安全注意事项

1. Personal Safety Equipment 个人安全设备

The workers must be fitted with the necessary personal safety equipment. Any safety equipment, whether collective or personal, must comply with existing standards and national regulations.

工作人员必须佩戴必要的个人安全设备。任何安全设备，不管是集体的还是个人的都必须符合现行标准和国家条例。

Hard hats for head protection.
安全帽，用于头部保护。

Safety shoes.
安全靴。

Safety glasses for eyes protection and adequate face shields for specific hazards such as: chipping, acid work, welding, molten metal, other risks of the same kind.
安全眼镜，用于眼部保护，具有足够的护面罩防止具体作业危害，例如：刨削、酸化作业、焊接、熔化金属以及同类的其他风险。

Safety mitten or **gloves** for hand protection.
安全连指手套或安全手套，用于手部保护。

Personnel when exposed to any hazard to ears must wear **ear protective devices**.
接触听力危害的人员必须戴上护耳装置。

Safety belt or **harness** complying with existing standards or (and) country regulations.
安全带，符合现行标准或(和)国家条例。

续表

 Protective mask with suitable filter when exposed to any hazard to the lungs caused by harmful fumes and gases spray painting, excess dust, and other risks of the same kind. 防护面罩，配有适当的过滤器，在接触有害烟气和气体喷涂、过量粉尘和同类的其他风险造成的肺部危害时佩戴。	 **Clean work clothes**: fireproof clothes are recommended when working in area where hazard of oxygen-rich atmosphere or presence of flammable product exist. Other adequate protective clothing will be prepared in case of exposure to specific hazards. 洁净工作服：建议在存在富氧大气危害的区域内作业时或有易燃产品存在时穿上防火服。在接触其他具体危害时应穿上适当的防护服。

2. Safety Measures 安全措施

(1) General Requirements 一般要求

■ Protective devices should be set up away from the dangerous zone but available to the watchs.

在远离危险区域但离看守人员很近的地方设置防护装置。

■ Special safety material: special extinguisher of automatic showers (carbon dioxide, water spray) for each kind of hazards, gas mask adapted to concerned gas, man-made mask etc.

特殊安全材料：适用于各种危险的自动喷淋专用灭火器（二氧化碳、喷水）、相关气体的防毒面具、自制面具等。

■ Informing and training the staff likely to work on polluted areas.

对可能在污染区域作业的人员进行通知和培训。

■ Use of adequate tools (anti-burning lamp, bronze tools 〈anti-sparks〉 free of oil and clean).

使用适当的工具（无油洁净的抗燃灯、青铜工具〈消火花〉）。

(2) General requirement on a plant 车间的一般要求

■ Clear signaling of hazardous zones (use barricades, sign posting).

Scene Five Safety Notes on Hoisting and Handling Equipment
情景五 设备吊装搬运安全注意事项

明确标出危险区域（利用路障、设置路标）。

■ Adequate bans (prohibited to smoke, to come in, to make fire).

张贴适当的禁令（禁止吸烟、禁止进入、禁止生火）。

■ Cleanliness (no materials on the ground, no oil pool, no oily rags).

清洁度（地上无材料、无油池、无破抹布）。

■ Availability of adequate help and safety equipments.

提供适当的帮助和安全设备。

(3) Areas considered as dangerous   危险区域划分

■ With difficult access (or difficult to evacuate) because of small exits (man hole) or exits position (height).

由于安全出口（检修孔）较小或安全出口位置（较高）的原因而难以进入（或难以撤离）的区域。

■ With closed area even when ventilated because accumulation in corners (analyzer panels, caskets, holes, sewers, ground with pit 〈gas heavier than air〉), ceilings (gas lighter than air) etc..

由于角落（分析仪表盘、容器、小孔、下水道、凹坑地面〈气体比空气重〉）和天花板（气体比空气轻）等位置易积聚，即使通风时也有封闭区的区域。

■ With cul-de-sac (sewers, pits, gutters, containers).

有凹陷的区域（下水道、凹坑、水槽、容器）。

■ In open air, near a gas source (nitrogen drying outlet) where atmosphere can be heavily polluted.

大气可能会被严重污染的露天区域和气源附近（氮气干燥出口）。

■ Places regarded as non dangerous but linked with dangerous places by piping (sewer).

认为没有危险但却与危险管道（下水道）位置相关的场所。

(4) Handling instructions   吊装说明

■ Handle with care. Operate according to the safety label on the equipment (Figure 5.1).

小心轻放。按设备上的安全标识（图5.1）进行操作。

■ Use only suitable lifting devices to ensure that no deformations, damages or other negative affects to surfaces and / or sealing areas can occur.

Figure 5.1 Equipment safety label

图5.1 设备安全标识

只能使用适当的起吊装置来确保表面和/或密封区域不会产生变形、损坏或其他负面影响。

■ Lifting shall be done by trained and qualified persons only (Figure 5.2).

起吊工作只能由经过培训的合格人员进行（图5.2）。

■ Lifting on nozzles and flanges of equipment or similar is not permitted. Use lifting straps for lifting.

禁止在设备的管嘴和法兰上或类似位置起吊。请使用起重带起吊。

■ Handling of equipment shall be executed in the way that any employee or third party will not be endangered.

搬运设备时不应使任何雇员或第三方受伤。

■ Lay down the BSU-cooling only at wooden pallets or similar wooden frames.

只能将燃烧室冷却辅助装置放在木托盘或类似木制框架上。

Scene Five Safety Notes on Hoisting and Handling Equipment
情景五 设备吊装搬运安全注意事项

Figure 5.2 Site safety announcement on hoisting and handling equipment
图 5.2 设备吊装搬运现场安全告知

Part II   Key Words and Expressions
第二部分 关键词汇

site engineer	工地工程师	workshop head	车间主任
director	厂长	chief of section	班组长
foreman	领工	worker	工人
staff member	职员	safety helmet/hard hats	安全帽
safety shoes	安全靴	safety glasses	安全眼镜

续表

safety mitten or gloves	安全连指手套或安全手套	ear protective devices	护耳装置
safety belt	安全带	protective masks	防护面罩
clean work clothes	洁净工作服	warning line	警戒线
electrode holder	焊钳	welding torch	焊炬
helmet shield	面罩	portable electrode heating box	手提式焊条加热箱
temperature measuring pen	测温笔	emergency gathering point	紧急集合点

Part Ⅲ　Typical Sentences
第三部分　典型单句

1	It is very simple and crude here.	这里很简陋。
2	May I introduce our chief engineer, Mr. Wang, to you?	我来介绍一下我们的总工程师,王先生。
3	We would like to listen to your opinions about our site work.	我们想听取你对我们现场工作的意见。
4	Pay attention to safety!	注意安全!
5	Put on your safety helmet, uniform, shoes, please.	请戴上安全帽、穿工装、工鞋。
6	Smoking and lighting fires are strictly forbidden here.	这里严禁烟火。
7	Look at the sign, danger keep out.	注意标牌,危险勿进。
8	Please pay attention to site warning line.	请注意现场警戒线。
9	I am sorry, do not touch this equipment, please!	很抱歉,请勿触动此设备!
10	All has gone well with our site work requirements.	一切均按照我们的现场工作要求进行。

Scene Five Safety Notes on Hoisting and Handling Equipment
情景五 设备吊装搬运安全注意事项

Part IV   Situational Dialogues
第四部分　实景开口说

Liu Jie：Mr. Brown, our work site is over there.
　　　　布朗先生，我们的施工现场在那边。

Brown：Oh, I see. Let's go.
　　　　哦，我看到了。我们过去吧。

Liu Jie：Welcome. The condition here is relatively primitive, I hope you don't mind.
　　　　欢迎。这里的条件有些简陋，希望你们不介意。

Brown：Oh, don't worry about this.
　　　　不会，我们不会介意。

Liu Jie：Thank you. This is our site engineer (director, workshop head, chief of section, foreman, worker, staff member), Mr. Chen, and chief engineer, Mr. Wang.
　　　　谢谢。这是我们的工地工程师（厂长、车间主任、班组长、领工、工人、职员），陈先生，以及我们的总工程师，王先生。

Chen & Wang：Nice to meet you. Mr. Brown.
　　　　　　　很高兴认识你，布朗先生。

Brown：Nice to meet you.
　　　　很高兴认识你们。

Wang：Here is our engineering office (drawing office, control room, laboratory, meeting room, rest room). Have a seat and rest a little bit, please.
　　　　这里是我们的工程技术室（图纸室、调度室、实验室、会议室、休息室）。请坐，休息一下。

Huebner/Brown：Thanks for your kindness. 感谢，费心了！

Wang：The two guys are toolpushers, Mr. Yang and Mr. Zhang.

这两位是带班队长，老杨、老张。

Huebner/Brown: Hi! Welcome to join us.
你好！欢迎你加入我们的行列！

Brown: We would like to listen to your opinions about our site work.
我们想听取你们对我们现场工作的意见。

Chen: Ok, here are some signs we need to know. Firstly, pay attention to safety, and put on your safety helmet, uniform and shoes. Secondly, smoking and lighting fires are strictly forbidden here. Thirdly, look at the sign, danger keep out.
好的。我们必须要记住一些标志。首先，注意安全，戴上安全帽、穿工装、工鞋。第二，这里严禁烟火。第三，注意标牌，危险勿近。

Liu Jie: And also, do not touch the equipment.
还有，请勿接触此设备。

Chen: In addition, we should wear this common tools bag. Pay attention to the site warning line, emergency gathering point.
此外还有，我们要佩戴好工具包。注意现场警戒线，紧急集合点。

Liu Jie: This sign means foreigners are not allowed in without permission.
这个标识的意思是未经允许，外国人不准入内。

Brown: I see.
明白了。

Huebner: A welder's kit contains electrode holder, welding torch, helmet shield, portable electrode heating box and temperature measuring pen.
一名焊工的成套工具包括焊钳、焊炬、面罩、手提式焊条加热箱和测温笔。

Chen Jun: Where's the *Record of Site Entry Registnation*?
入场登记记录在哪儿？

Wang: In the drawer of the table.
在桌子抽屉里。

Scene Five Safety Notes on Hoisting and Handling Equipment
情景五 设备吊装搬运安全注意事项

Huebner, Brown: Ok, let's fill out the *Record of Site Entry Registnation*.

好，我们填写入场登记记录。

Wang: Everything goes well following our requirements.

一切都按照我们的要求进行着。

Brown: Great.

太好了！

Part V Exercises
第五部分 练习活动

Task 1: Study the following words and match the words in left column with corresponding Chinese phrases or terminologies in the right column.

____ 1. hoisting and handling	a. 吊车
____ 2. safety measures	b. 吊装说明
____ 3. warning of hand injury	c. 热表面警告
____ 4. ear protective devices	d. 个人安全设备
____ 5. crane	e. 安全带
____ 6. safety belt	f. 吊装
____ 7. personal safety equipment	g. 安全措施
____ 8. safety sign	h. 护耳装置
____ 9. warning of hot surface	i. 伤手警告
____ 10. handling instructions	j. 安全标志

Task 2: Make sure you know how to spell and pronounce all the words above. Then write a sentence for each word. The sentence must show you understand the meaning of the word.

_____

_____

_____

Task 3: Translate the following sentences into Chinese/English.

1. China National Chemical Construction Corporation (CNCCC) contracts for domestic and overseas chemical projects.

2. Are you the Seller's Representative on the job site?

3. The project team normally consists of project engineer, design engineer, schedule engineer, and various specialists.

4. 这个项目的合同号是 NP-S-01/C。

5. 买方是中国技术进口总公司。

6. 我是买方的总代表。

Task 4: Role Play.
Follow the example of the conversation above. Work together as a group to create an episode on this task theme. And then use role play to practice your English communication skills on safety notes on hoisting and handling equipment.

Scene Five Safety Notes on Hoisting and Handling Equipment
情景五 设备吊装搬运安全注意事项

情景模拟演练任务书

班级	
任务主题	Safety Notes on Hoisting and Handling Equipment 设备吊装搬运安全注意事项
组名、成员	
演练内容	Use the expressions about this task with some communication skills(e. g. describing) to create a dialogue, an episode etc. 运用与本任务主题有关的语言表达和交流技能(如描述)创设情景会话或情景剧。
演练目的	1. master the expressions about this task. 掌握本任务的表达方式。 2. master the communication skills in English. 掌握英语交流技能。 3. put them into application. 把所学付诸于实践应用。
小组成员承担的 任务及角色	
任务完成情况 及步骤	
小组任务完成 综合体会(综合 成员个人体会)	

情景模拟演练成果展示评分标准

考核内容	分值	得分	备注
一、设计思路表达是否切合任务主题	20		
二、内容是否积极、健康、向上	20		
三、在角色表演过程中是否投入、认真、代入感强	20		
四、编选的情景模拟要自然，符合基本的英语表达习惯，是否掌握正确运用交际用语的能力	20		
五、语音语调、手势表情、即兴反应	10		
六、小组成员协作与团队精神体现	10		

评价 \ 得分	小组最终得分	小组成员最终得分
小组互评（包括所有成员）		
教师点评		

Scene Five Safety Notes on Hoisting and Handling Equipment
情景五 设备吊装搬运安全注意事项

Attachment: Episode

附： 情景剧稿

Scene Six 情景六

Discussing the Technical Documents, Drawings and Instructions with Foreigners before Lifting
设备吊装前与外方人员讨论技术资料、图纸和说明书

Part I Background Information
第一部分 情景背景知识

1. Method and Sequence for Installation of Process Equipment

工艺设备安装方法和安装原则

Installation Method: equipment installed at spacious area shall adopt the method of centralized setting in place by crane or large professional crane. Equipment unable to be set in place directly by crane can be located in place by using chain blocks plus rolling bars. But the ground for civil construction must be effectively protected, paved soft rubber sheet and hard fiber board. Equipment with smaller volume and weight less than 3 tons will be located in place by manual hydraulic forklift for safety.

安装方法：宽敞区域的设备采用汽车吊机或大型专业吊机集中就位方法，不能用吊机直接就位的设备可以用手拉葫芦＋滚杠辅助就位，但必须对土建地面进行有效保护，铺上软质橡胶板和硬质纤维板。体积较小和重量小于3吨的设备采用手动液压叉车进行安全就位。

Installation Principle: lifting of equipment will be done in accordance with the sequence such as lower equipment before upper equipment, larger equipment before smaller equipment, inside equipment before outside equipment, heavier equipment before lighter equipment to save energy and reduce consumption. Our company will choose technicians and construction teams with rich installation experiences, apply advanced technique and ensure construction quality and schedule to satisfy the requirement of equipment installation.

安装原则：设备吊装先下后上、先大后小、先里后外、先重后轻，合理安排顺序，节能降耗；我公司将选择安装经验丰富的技术人员和施工班组，采用先进技术确保设备安装的施工质量和施工进度。

2. Warning 警告

◆ Do not lift the unit by lifting eyes or similar provisions provided on

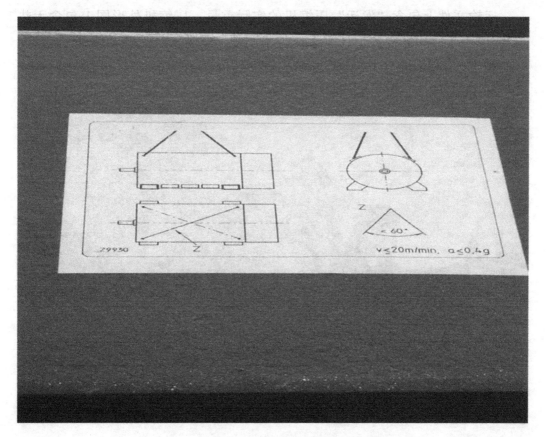

Figure 6.1　Imported equipment lifting nameplate
图 6.1　进口设备吊装铭牌

various package components. These are only suitable for lifting the individual component.

禁止使用各包装组件上的吊耳或类似部位进行机组起吊。此类部位只适合起吊单个部件。

◆ All boxed units have lifting points printed at the appropriate locations. Select crane capacities and lifting accessories accordingly. The imported equipment lifting nameplate is shown in Figure 6.1.

所有包装的组件均有吊点并印在了合适的位置。请对应选择起吊能力和起吊配件。进口设备吊装铭牌如图6.1所示。

◆ Actual weights of the "as shipped" compressor are shown on the shipping documents. Estimated weights of the compressor without skid and/or box is shown on the compressor outline drawing. Technical drawings, driver weights are shown on the nameplate.

运输文件上包含"发运"压缩机的实际重量。压缩机外形图上包含无垫木和/或箱子的压缩机的估计重量。技术图纸、驱动重量见驱动铭牌。

◆ Before lifting, pay attention to read the drawing of equipment assembly (Figure 6.2) and discuss lifting drawings and instructions (Figure

Figure 6.2　Reading the drawing of equipment assembly

图6.2　阅读设备组装图纸

Scene Six Discussing the Technical Documents, Drawings and Instructions with Foreigners before Lifting 情景六 设备吊装前与外方人员讨论技术资料、图纸和说明书

6.3）。

　　实施吊装前，要注意阅读设备组装图纸（图 6.2）和讨论吊装图纸资料与说明书（图 6.3）。

Figure 6.3 Discussing lifting drawings and instructions

图 6.3 讨论吊装图纸资料与说明书

Example 1

案例 1

Lifting the Gearbox Assembly

吊装齿轮箱组件

Carefully secure the lifting chains around the gearbox lifting ears to support the weight of the assembly. Be sure these chains are tight with no slack.

仔细将吊链固定在齿轮箱的吊耳周围，以承受组件重量。确保这些吊链张紧无松弛。

For compressors with two split lines (cover rotor), secure the adjustable safety chains to the eyebolts in the gearbox cover to prevent tipping of the top heavy assembly. Attach the safety chains snugly to the gearbox cover eyebolts. The gearbox assembly should not be lifted only by the eyebolts in the gearbox cover, or the lifting eye on the cover-mounted scroll.

对于带两条分模线的压缩机（盖住转子），将可调节安全链固定到齿轮箱盖内的吊环螺栓上，以防顶部重型组件倾翻。将安全键紧紧地连接到变速箱盖吊环螺栓上。任何时候都不能只采用齿轮箱盖内的吊环螺栓或罩盖上安装的滚轴上的吊眼来吊装齿轮箱组件。

In some cases, with large scrolls only on one side of the gearbox, it may be necessary to secure chains to the inlet (s) of the larger scrolls to prevent the box from tipping (Figure 6.4).

在一些情况下，由于大型滚轴仅位于齿轮箱的一侧，因此可能需要将吊链固定在较大的滚轴上，以防箱子倾翻（如图6.4所示）。

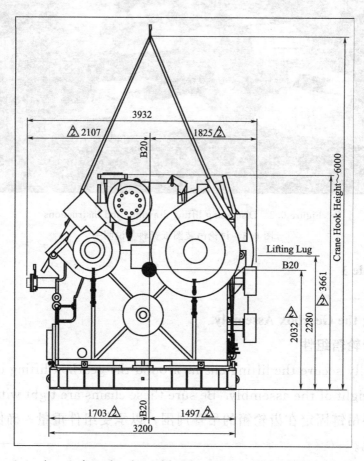

Figure 6.4 A typical gearbox being lifted by the ears with stabilizing chains to the cover to prevent tipping

图6.4 将吊链固定到罩盖上以防箱子倾翻，利用吊耳进行吊装的典型齿轮箱

If the gearbox is enclosed in shrink warp, open appropriate zipper panels to route chains in correct locations. Shrink warp integrity must be maintained, and minimize stress of the wrapping by the rigging wherever possible. Ensure that zipper covers are fully closed and shrink wrap is intact after lifting gearbox.

如果齿轮箱放在热缩塑料包内，则打开适当拉锁板，将吊链放在正确位置。必须保持热缩塑料包完整，尽可能减少索具对热缩塑料包造成的压力。吊装齿轮箱后，确保拉锁盖完全关闭，热缩塑料包完好无损。

CAUTION   注意

Do not lift the unit by means of the eyebolts on the gearbox or recesses on the driver. These are for the maintenance of specific components only.

不得利用齿轮箱上的吊环螺栓或驱动器上的凹槽对装置进行吊装。这些部位仅用于具体部件的维护。

Example 2

案例 2

Lifting compressor without motor is recommended.

建议压缩机进行无电动机起吊。

Compressor units have four (4) lifting points as shown in Figure 6.5. Do not attempt to lift unit with any other procedures or points than those specified in the figure. A safety hazard or unit damage may occur otherwise.

如图 6.5 所示，压缩机机组有四个（4）吊点。不要试图按照图中标注之外的程序或吊点起吊设备。否则可能造成安全事故或设备损坏。

Adjust length of cables so crane hook is positioned above center of gravity of the compressor, to insure compressor remains level while lifting.

调节缆绳的长度，使起重机吊钩定位在压缩机重心之上，以确保压缩机吊起时保持水平。

Dimensions x_1, x_2, and x_3 should be equal to or less than dimensions y_1, y_2, and y_3 respectively to maintain stability while lifting.

尺寸 x_1、x_2 和 x_3 应该分别等于或小于尺寸 y_1、y_2、y_3，以在起吊时保持稳定。

Spreader bars must be sufficiently strong to resist all buckling loads. Secure cables at points indicated by "+" to prevent movement while lift-

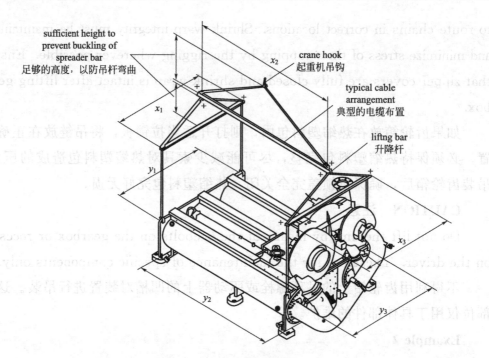

Figure 6.5 Suggested lifting arrangement of compressor

图 6.5 压缩机建议起吊布置

ing. During lifting operation do not touch inlet and discharge piping and/or any other component with lifting cables/chains.

撑杆必须足够结实，可以抵挡所有屈曲载荷。在标有"+"的位置固定缆绳，防止起吊时晃动。起重作业期间切勿触摸进气和出气管道和/或任何其他带有起吊缆绳/链条的组件。

Part Ⅱ Key Words and Expressions
第二部分 关键词汇

norm	规范	standard	标准
rules of operation	操作规程	erection	安装
alignment	对中	testing	试验
plot plan	平面布置图	general layout	总平面

Scene Six Discussing the Technical Documents, Drawings and Instructions with Foreigners before Lifting 情景六 设备吊装前与外方人员讨论技术资料、图纸和说明书

续表

general arrangement	总布置	detail	细部
section	剖面	flow sheet	流程
P&ID	带仪表控制点的管道	assembly	装配
civil	土建	electric	电气
control and instrumentation	自控和仪表	projection	投影
piping	配管	border lines	边框线
visible lines	实线	invisible lines	虚线
break lines	断裂线	phantom lines	假想线
abbreviation	缩写	mark	标记
symbol	符号		

Part Ⅲ Typical Sentences
第三部分 典型单句

1	According to the technical standard, the erection work is under way.	安装工作正在根据技术标准进行。
2	This is a plot plan of erection drawing.	这是一张安装平面布置图。
3	That is a general view.	那是全视图。
4	Please explain the meaning of this abbreviation on the drawing.	请解释图上这个缩写的意义。
5	We comply with and carry out the DIN standard in this project.	在这个工程中我们遵守并执行联邦德国标准 DIN。
6	It maybe not quite sure.	可能不太确切。
7	A working drawing must be clear and complete.	工作图必须简明完整。
8	Please have a look at the drawing.	请看这张图。
9	We shall do our best.	我们尽力做好。

Part Ⅳ Situational Dialogues
第四部分 实景开口说

Liu Jie: Mr. Brown, what is this?

  布朗先生,这是什么?

Brown: This is a general (front, rear, side, left, right, top, vertical, bottom, elevation, auxiliary, cut-away, birds eye) view of a plot plan of erection (general layout, general arrangement, detail, section, flow sheet, P&ID, assembly, civil, electrical, control and instrumentation, projection, piping) drawing. We will complete our work following this drawing.

  这是一张全视角(前视、后视、侧视、左视、右视、顶视、俯视、底视、立视、辅助、内部剖视、鸟瞰)安装平面布置(总平面、总布置、细部、剖面、流程、管道及仪表流程、装配、土建、电气、自控和仪表、投影、配管)图。我们将按照这张图纸完成这项工作。

Huebner: The erection (alignment, testing) work is under way according to the technical standard (norm, rules of operation).

  安装(校准、实验)工作正在根据技术标准(规范、操作规程)进行。

Liu Jie: How many drawings are there in the set?

  这套图纸有几张?

Brown: This is just a copy, only one.

  这是复制的副本图,就一张。

Huebner: The information to be placed in each title block of a drawing include: drawing number, drawing size, scale, weight,

Scene Six Discussing the Technical Documents, Drawings and Instructions with Foreigners before Lifting 情景六 设备吊装前与外方人员讨论技术资料、图纸和说明书

  sheet number and number of sheets, drawing title and signatures of persons preparing, checking and approving the drawing.

  每张图纸的图标栏内容包括：图号、图纸尺寸、比例、重量、张号和张数、图题，以及制图、校对、批准人的签字。

Brown: There are various lines on the drawing such as: border lines, visible lines, invisible lines, break lines, phantom lines.

  图上有各种形式的线条，诸如：边框线、实线、虚线、断裂线、假想线。

Liu Jie: Could you please explain the meaning of these abbreviations (marks, symbols)?

  您能解释一下这个缩写（标记、符号）表示什么意思呢？

Brown: Yes, this means we comply with and carry out the DIN (BS, NF, JIS) standard in this project.

  好的。它的意思是在这个工程中，我们要遵守并执行联邦德国标准DIN（英国标准、法国标准、日本工业标准）。

Liu Jie: I see. Can we have this drawing?

  明白了。这图纸可以留给我们吗？

Brown: Sure, you can. But this is just a translated version, and it maybe not quite correct.

  当然可以。只不过这是翻译过来的版本，可能有一些内容不太准确。

Liu Jie: Emm... I found a mistake in this translation.

  嗯……我看到了一个错误。

Huebner: You are smart. As we all know, a working drawing must be clear and complete.

  你真聪明！我们都知道，工作图必须简明完整。

Liu Jie: Please check the drawings. They would be able to do this after some training.

  请看这张图。经过一些训练，他们就能胜任这项工作。

Huebner: We will do our best.
我们尽力做好。

Liu Jie: Thank you.
谢谢！

Part V Exercises
第五部分 练习活动

Task 1: Study the following words and match the words in left column with corresponding Chinese phrases or terminologies in the right column.

____ 1. technical documents   a. 机电一体化系统设计
____ 2. mechanical-electrical integration system design   b. 安装方法
____ 3. lifting compressor without motor   c. 规范
____ 4. norm   d. 标记
____ 5. plot plan   e. 操作规程
____ 6. Drawings and Instructions   f. 技术资料
____ 7. lifting ears   g. 吊耳
____ 8. rules of operation   h. 平面布置图
____ 9. mark   i. 无电动机起吊
____ 10. installation method   j. 工艺设备
____ 11. kinematic sketch of mechanism   k. 机构组成
____ 12. process equipment   l. 偏（心）距
____ 13. block diagram   m. 机构运动简图
____ 14. constitution of mechanism   n. 图纸和说明书
____ 15. offset distance   o. 框图

Task 2: Make sure you know how to spell and pronounce all the words above. Then write a sentence for each word. The sentence must show you understand the meaning of the word.

Scene Six Discussing the Technical Documents, Drawings and Instructions with Foreigners before Lifting 情景六 设备吊装前与外方人员讨论技术资料、图纸和说明书

Task 3: Translate the following sentences into Chinese/English.

1. We completed this task according to the drawing number SD-102.

2. What is the edition of this drawing?

3. We have not received this drawing (instruction book, operation manual), please help us to get it.

4. 我负责这个项目（区域）的技术（检查、质量控制）工作。

5. 请把那些图纸带给我们。

6. 可否将资料（手册、小册）递给我？

Task 4: Role Play.
Follow the example of the conversation above. Work together as a group to create an episode on this task theme. And then use role play to practice your English communication skills on discussing the technical documents, drawings and instructions with foreigners before lifting.

情景模拟演练任务书

班级	
任务主题	Discussing the Technical Documents, Drawings and Instructions with Foreigners before Lifting 设备吊装前与外方人员讨论技术资料、图纸和说明书
组名、成员	
演练内容	Use the expressions about this task with some communication skills (e.g. describing) to create a dialogue, an episode etc. 运用与本任务主题有关的语言表达和交流技能(如描述)创设情景会话或情景剧等。
演练目的	1. master the expressions about this task. 掌握本任务的表达方式。 2. master the communication skills in English. 掌握英语交流技能。 3. put them into application. 把所学付诸于实践应用。
小组成员承担的任务及角色	
任务完成情况及步骤	
小组任务完成综合体会(综合成员个人体会)	

情景模拟演练成果展示评分标准

考核内容	分值	得分	备注
一、设计思路表达是否切合任务主题	20		
二、内容是否积极、健康、向上	20		
三、在角色表演过程中是否投入、认真、代入感强	20		
四、编选的情景模拟要自然，符合基本的英语表达习惯，是否掌握正确运用交际用语的能力	20		
五、语音语调、手势表情、即兴反应	10		
六、小组成员协作与团队精神体现	10		

评价 \ 得分	小组最终得分	小组成员最终得分
小组互评（包括所有成员）		
教师点评		

Attachment: Episode

附： 情景剧稿

Scene Seven 情景七

Hoisting Imported Equipment with Foreigners
与外方人员一起吊装进口设备

Part I Background Information
第一部分 情景背景知识

1. Equipment Locating 设备就位

Before locating equipment, first check the nozzle orientation, nozzle size and relative nozzle position of the equipment and identify nozzle number. Equipment center location shall be marked on the base. The equipment shall be located and aligned with the centerline of the equipment base. Then apply reasonable and reliable method to lift the equipment and set in place. Benchmark for static equipment adjustment and measurement shall be specified as follows.

设备就位前,首先核对设备的管口方位、管口尺寸、管口相对位置,并标注管号。在设备底座上标注设备中心方位,就位时应与设备基础方位中心线对齐,然后选择合理可靠的方法将设备吊装就位。静设备调整和测量的基准规定如下。

(1) Bottom elevation of the equipment bearing shall be based on the elevation benchmark line on the foundation. 设备支承的底面标高应以基础上标高基准线为基准。

(2) Equipment centerline location shall be based on the centerline location on the foundation. 设备中心线位置应以基础上中心线位置为基准。

(3) Orientation of vertical equipment shall be based on the centerline closest to the equipment on the foundation. 立式设备的方位应以基础上距离设备最近的中心线为基准。

(4) Plumbness of vertical equipment shall be based on the measuring points at both ends of equipment. 立式设备的铅垂度应以设备两端部的测点为基准。

(5) Levelness of horizontal equipment shall be generally based on the centerline of the equipment. 卧式设备的水平度一般应以设备的中心线为基准。

For allowable deviation of equipment installation please refer to the following table.

设备安装允许偏差，见下表：

Allowable error of equipment installation 设备安装允许偏差

Item 项目 \ Type 型式	Vertical Equipment 立式设备	Horizontal Equipment 卧式设备
	Allowable Deviation 允许偏差	
Centerline Position 中心线位置	±5mm	
Elevation 标高	±5mm	
Orientation 方位	Measurement along the base ring circumference not over 15mm 沿底座环圆周测量不超过15mm	
Plumbness 垂直度	$h/1000$, but not over 25mm 为 $h/1000$,但不超过25mm	
Levelness 水平度		Axial $L/1000$, Radial $2D/1000$ 轴向为 $L/1000$,径向为 $2D/1000$

Note：L—distance between two supporting seats of horizontal equipment；D—outside diameter of equipment；h—distance between two-end measuring points of vertical equipment. The specific location of sizes are shown in Figure 7.1.

注：L—卧式设备两支座间距离；D—设备外径；h—立式设备两端测点距离。尺寸的具体位置见图7.1。

2. Necessary devices/tools for lifting 吊装所需装置/工具

Figure 7.1 to Figure 7.5 show the important tools for lifting, and Figure 7.6 to Figure 7.8 show the application of lifting tools on site.

Scene Seven Hoisting Imported Equipment with Foreigners
情景七 与外方人员一起吊装进口设备

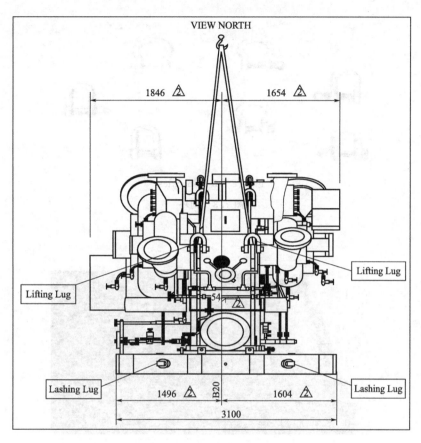

Figure 7.1 Lifting view north
图 7.1 吊装端视图

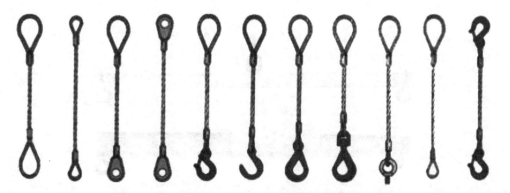

Figure 7.2 Steel cables for lifting
图 7.2 起吊钢索

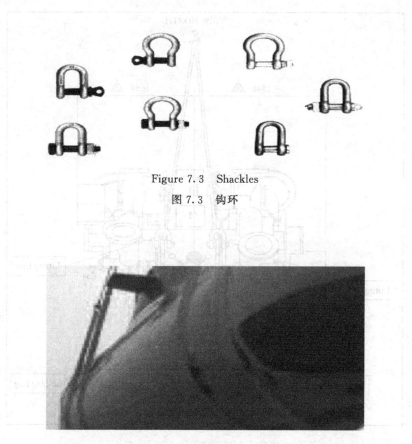

Figure 7.3 Shackles
图 7.3 钩环

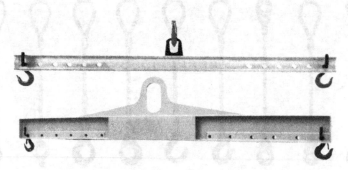

Figure 7.4 Lifting straps
图 7.4 起重带

Figure 7.5 Lifting beam
图 7.5 吊梁

Scene Seven Hoisting Imported Equipment with Foreigners
情景七 与外方人员一起吊装进口设备

Figure 7.6   Hoisting with hanging chains and lifting beams on site
图 7.6   使用吊链和吊梁吊装现场图

Figure 7.7   Hoisting with crawler crane on site A
图 7.7   在 A 现场使用履带起重机吊装

Figure 7.8　Hoisting with crawler crane on site B
图 7.8　在 B 现场使用履带起重机吊装

图 7.1～图 7.5 为重要的吊装工具，图 7.6～图 7.8 为吊装工具在施工现场的应用。

Part Ⅱ　Key Words and Expressions
第二部分　关键词汇

truck crane	汽车起重机	gantry crane	龙门起重机
tower crane	塔式起重机	mobile slewing crane	悬臂汽车起重机
bridge crane	桥式起重机	crawler crane	履带式起重机

Scene Seven Hoisting Imported Equipment with Foreigners
情景七 与外方人员一起吊装进口设备

续表

gin pole	起重抱杆	girder pole	主架抱杆
guy derrick	转盘抱杆	hammer	铁锤
hacksaw	钢锯	file	锉刀
scraper	刮刀	chisel	凿子
socket wrench	套筒扳手	hook spanner	钩扳手
adjustable wrench	活动扳手	pipe wrench	管扳手
ratchet wrench	棘轮扳手	open end wrench	开口扳手
screw driver	螺丝刀	hand vice	钳子
pliers	钳子	pocket knife	小折刀
check-over	全面检查	repairing	修理
overhaul	大修	welder	焊工

Part III Typical Sentences
第三部分 典型单句

1	We have all kinds of construction machinery on the site.	我们有各种施工机械在现场。
2	The truck crane can lift a weight of 15 tons.	这台汽车起重机能吊起15吨的重物。
3	Pass me a hammer.	给我拿一把铁锤。
4	Lifting ear is here.	吊耳是在这个位置。
5	Thanks for your reminding.	谢谢你的提醒。
6	All has gone well.	一切准备就绪。
7	Danger! Look out! Get out of the way.	危险！当心！快躲开！
8	Look at the sign, danger keep out.	注意标牌，危险勿进。
9	What happened?	怎么了？
10	Would you like to talk to the welder?	你要和焊工谈谈吗？
11	Something must be wrong with it.	这一定出问题了。

Part Ⅳ　Situational Dialogues
第四部分　实景开口说

Li Lei: Nice to meet you, Mr. Brown. I am Li Lei.
　　　你好，布朗先生，我叫李磊。

Brown: Nice to meet you, Mr. Li.
　　　你好李先生。

Li Lei: On this work site, you can see various machineries of construction.
　　　在这个工地上，我们有各种施工机械。

Brown: Oh great. Can you give me some introductions?
　　　哦太好了，你能给我介绍一下吗？

Chen Jun: Sure. This is a truck crane (gantry crane, tower crane, mobile slewing crane, bridge crane, crawler crane). It can lift a weight of 15 tons.
　　　好的。这是一台汽车起重机（龙门起重机、塔式起重机、悬臂汽车起重机、桥式起重机、履带式起重机）。能吊起15吨的重物。

Li Lei: The hoisting capacity of that gin pole is 60 tons.
　　　那个起重抱杆的起重力是60吨。

Huebner: Is this "LIEBRHERR" (HUANGHE, NISSAN, TOYOTA, ISUZU, MITSUBISHI, HITACHI, GMC)?
　　　这台吊机是利勃海尔（黄河、尼桑、丰田、五十铃、三菱、日立、通用）吗？

Chen Jun: Yes! That's it! Load at the hook, 206t.
　　　是的，就是它！吊钩载重量为206吨。

Brown: Pass me a hammer (hacksaw, file, scraper, chisel, socket

Scene Seven Hoisting Imported Equipment with Foreigners
情景七 与外方人员一起吊装进口设备

    wrench, hook spanner, adjustable wrench, pipe wrench, ratchet wrench, open end wrench, screw driver, hand vice, pliers, pocket knife).

    给我拿一把手锤（钢锯、锉刀、刮刀、凿子、套筒扳手、钩扳手、活动扳手、管扳手、棘轮扳手、开口扳手、螺丝刀、手钳、钳子、小折刀）。

Li Lei：Here you are.

    给您！

Brown：We must study the caution instruction before lifting this Compressor-Intercooler-Base Assembly.

    在吊装压缩机-中冷器-基座组件之前，我们必须先学习吊装注意事项。

Huebner：Lifting ear is here.

    吊耳是在这个位置。

Li Lei：Ok, I see.

    哦，看到了。

Brown：This Compressor-Intercooler-Base Assembly is being lifted by the ears with stabilizing chains to the cover to prevent tipping.

    这个压缩机-中冷器-基座组件吊装时利用吊耳进行吊装，吊链固定到罩盖上以防倾翻。

Li Lei：Oh, thanks for your reminding.

    谢谢你的提醒。

Chen Jun：All has gone well.

    一切准备就绪。

Li Lei：Alright!

    哦，知道了。

Brown：One, two, three, ready, go. It is normal. It is clear. It is correct. It is all right.

    一、二、三，准备，起吊！正常！清楚！正确！良好！

Huebner：Danger! Look out! Get out of the way. Look at the sign, danger keep out.

危险！当心！快躲开！注意标牌，危险勿进。

Brown: To maintain its efficiency, the machinery needs a regular service (check-over, repairing, overhaul).

这台机械需要进行一次定期保养（全面检查、修理、大修），以维持其工作效率。

Chen Jun: What happened? Would you like to talk to the welder (inspector)?

怎么了？你要和焊工（检查员）谈谈吗？

Brown: Yes, something must be wrong with it. I'd like to talk with them about this problem.

是的，出问题了，我想针对这个问题和他谈一下。

Li Lei: Let me get him. Hi... Excuse me, Mr. Wang.

我去找他。嗨……，打扰了，王先生。

Wang: Yes? Why are you so hurry?

怎么了？怎么这么着急？

Li Lei: Mr. Brown would have a talk with you about some problems he founded during the inspection on the nozzle orientation.

布朗先生想和你谈谈他在检查管口方位过程中发现的一些问题。

Wang: Sure. I will be there soon.

好的。我立马就去。

Brown: Hi, Wang. Do you know the allowable deviation of this nozzle orientation?

王先生，你来了。你知道这个管口方位的允许偏差吗？

Wang: Exactly, the measurement along the base ring circumference should not be over 15mm.

准确的来说，沿底座环圆周测量不超过15mm。

Brown: Got it. All right.

好的，明白了。

Li Lei: I learn from you guys.

我也懂了。

Scene Seven Hoisting Imported Equipment with Foreigners
情景七　与外方人员一起吊装进口设备

Part V Exercises
第五部分　练习活动

Task 1：Study the following words and match the words in left column with corresponding Chinese phrases or terminologies in the right column.

____ 1. static equipment	a. 立式设备
____ 2. steel cables for lifting	b. 垂直度
____ 3. lifting view north	c. 龙门吊
____ 4. levelness	d. 允许偏差
____ 5. elevation	e. 吊装端视图
____ 6. gantry crane	f. 水平度
____ 7. allowable error	g. 标高
____ 8. plumbness	h. 起吊钢索
____ 9. mobile slewing crane	i. 静设备
____ 10. vertical equipment	j. 悬臂汽车吊
____ 11. equipment locating	k. 端面重合度
____ 12. transverse contact ratio	l. 径向
____ 13. load	m. 吊梁
____ 14. radial direction	n. 设备就位
____ 15. lifting beam	o. 载荷

Task 2：Try your best to use the real work situation to list the specification of benchmark briefly for static equipment adjustment and measurement.

_____
_____
_____
_____

Task 3: Translate the following sentences into Chinese/English.

1. There is a temporary facility for the hoisting site.

2. All has gone well with hoisting site work plan.

3. Hoisting mechanism must be lubricated periodically.

4. 吊装一切正常……清楚……正确……良好!

5. 请告诉我如何操作这台吊臂。

6. 这台起重转盘抱杆的起重能力为60吨。

Task 4: Role Play.

Follow the example of the conversation above. Work together as a group to create an episode on this task theme. And then use role play to practice your English communication skills on hoisting imported equipment with foreigners.

Scene Seven Hoisting Imported Equipment with Foreigners
情景七 与外方人员一起吊装进口设备

情景模拟演练任务书

班级	
任务主题	Hoisting Imported Equipment with Foreigners 与外方人员一起吊装设备
组名、成员	
演练内容	Use the expressions about this task with some communication skills(e. g. describing) to create a dialogue, an episode etc. 运用与本任务主题有关的语言表达和交流技能（如描述）创设情景会话或情景剧等。
演练目的	1. master the expressions about this task. 掌握本任务的表达方式。 2. master the communication skills in English. 掌握英语交流技能。 3. put them into application. 把所学付诸于实践应用。
小组成员承担的任务及角色	
任务完成情况及步骤	
小组任务完成综合体会（综合成员个人体会）	

情景模拟演练成果展示评分标准

考核内容	分值	得分	备注
一、设计思路表达是否切合任务主题	20		
二、内容是否积极、健康、向上	20		
三、在角色表演过程中是否投入、认真、代入感强	20		
四、编选的情景模拟要自然，符合基本的英语表达习惯，是否掌握正确运用交际用语的能力	20		
五、语音语调、手势表情、即兴反应	10		
六、小组成员协作与团队精神体现	10		

评价＼得分	小组最终得分	小组成员最终得分
小组互评（包括所有成员）		
教师点评		

Attachment: Episode

附：情景剧稿

Scene Eight 情景八

Installing, Testing, Starting-up, Disassembling and Maintaining with Foreigners
与外方人员安装、调试、开车、拆检和维护

Part I Background Information
第一部分 情景背景知识

1. Machines Installation 机器安装

Good planning and preparation result in efficient and accurate installation. The machine is carefully lifted and placed on the foundation. A rough horizontal alignment is made with the aid of the previously installed steel wire and the marking of the axial location. A vertical alignment is made with the leveling screws. Required positioning accuracy is within 2mm. On-site equipment installation is shown in Figure 8.1.

良好的规划和准备可以确保安装的简易性和正确性。小心地吊装该机器并且将其放置在基础上。在先前安装的钢丝和轴位置标记的援助下，进行粗略的水平校准。用水平螺钉进行垂直校准。所要求的定位精度在2毫米内。现场设备安装情况如图8.1所示。

Alignment and level of the equipment should be in accordance with the following specifications. 设备的找正、找平应符合下列规定。

Scene Eight Installing, Testing, Starting-up, Disassembling and Maintaining with Foreigners 情景八 与外方人员安装、调试、开车、拆检和维护

Figure 8.1   On-site equipment installation

图 8.1   设备安装

★ Alignment and level work should be done on the basis of two and more than two directions at the right angle on the same plane.

找正和找平应在同一平面内互成直角的两个或两个以上的方向进行。

★ For the equipment over 20m in height, the adjustment and measurement of plumbness should prevent performance from the condition of direct sunlight at one side and the wind stronger than 4 scale.

高度超过 20 米的立式设备，其铅垂度调整和测量工作应避免在一侧受强阳光照射及风力大于 4 级的条件下进行。

★ Alignment and level work of the equipment should be adjusted with sizing block as required. No adjustment is allowed by using the method of fastening or unfastening anchor bolts and local pressurization.

设备的找正和找平应根据要求进行垫铁调整，不得用紧固或放松地脚螺栓及局部加压等方法进行调整。

★ Fastening of anchor bolts should be even and symmetric.

地脚螺栓的紧固应均匀对称地进行。

2. Alignment 找正

To ensure a long and satisfactory lifetime of both the driving and the driven machine, the machines need to be properly aligned to each other.

为了确保驱动机和从动机能够达到令人满意的长使用寿命,需要正确地找正对齐。

Before the alignment procedure is started, the coupling half have to be installed for coupling halves of the driving and the driven machines. The measurement of run-out of the coupling half shown in Figure 8.2.

为了耦合驱动机和从动机的一半,在校准过程开始之前,必须安装半联轴器。半联轴器跳动的测量示意图见图8.2。

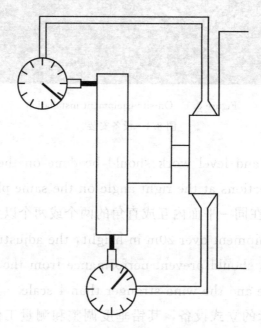

Figure 8.2　Measuring the run-out of the coupling half

图 8.2　测量半联轴器的跳动

3. Rough levelling 粗调平

Check the machine's position from vertical, horizontal and axial direction. Make adjustments accordingly by placing shims under the four feet. The horizontal level of the machine is checked with a spirit level or machine foot, as the Figure 8.3.

检查该机器的垂直、水平和轴向的水平位置。通过将垫片放置在四个机脚的下方来进行相应的调整。用水平仪或机脚检查该机器的水平位置，如图8.3。

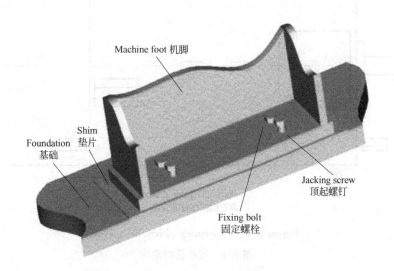

Figure 8.3 Vertical positioning of machine foot
图 8.3 机脚的垂直定位

4. Rough adjustment 粗调整

To facilitate the alignment in axial and transversal directions, fix the bracket plates with adjusting screws at the corners (Figure 8.4).

为了便于在轴向和横向方向上校准，用调整螺钉将支架板固定在拐角处（图8.4）。

Bracket plates are placed against the foundation edge and tied down with expansion bolts (Figure 8.5). Use the adjusting screws to move the machine until the shaft centerline and the driven machine centerline are aligned roughly and the desired distance between the coupling halves is reached. Leave all adjusting screws only lightly tightened.

将支架板放置在基础边缘并且用膨胀螺栓固定（图8.5）。使用调整螺钉移动该机器，直到轴中心线大概与从动机中心线相校准并且达到半联轴器之间的期望距离。仅保留所有轻轻拧紧的调节螺钉。

NOTE: the picture of Mounting of the bracket plate shows bracket plate mounted on concrete

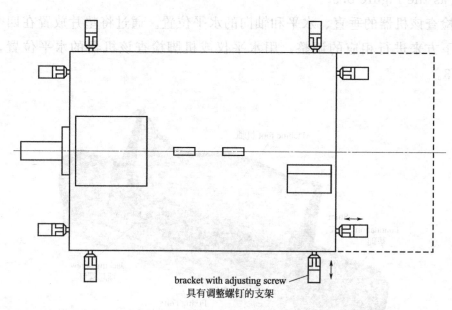

Figure 8.4　Positioning of bracket plates

图 8.4　支架板的定位

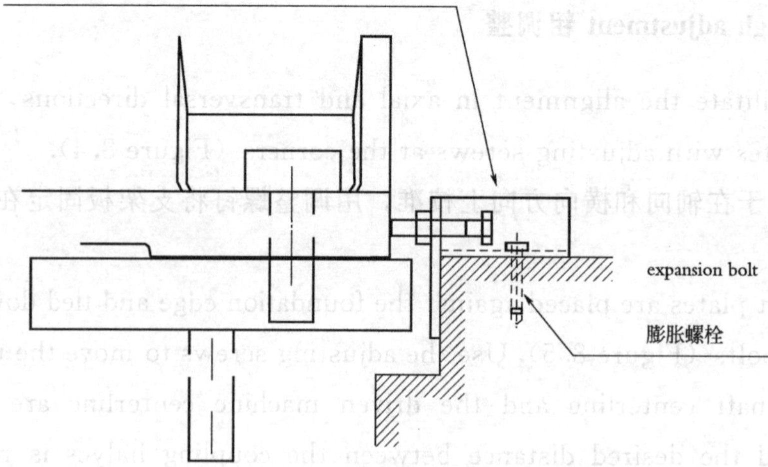

Figure 8.5　Mounting of the bracket plate

图 8.5　支架板的安装

foundation, placed similar bracket plate on steel foundation.

注释：支架板的安装图显示出了将支架板安装到混凝土基础之上，将类似的支架板放置在钢基础之上。

5. Final alignment 最终找正

In the following, the final alignment is made with dial gauges (Figure 8.6).

下面还应该用千分表进行最终找正（图8.6）。

NOTE: The final alignment measurements should always be recorded for future reference.
注释：应该始终记录最终校准测量值，以备将来参考。

Figure 8.6　Using dial gauges for final alignment
图8.6　使用千分表最终找正

6. Grouting 灌浆

The grouting of the machine into the foundation is a very important part of the installation. Cracks in the grouting compounds or a poor attachment to the concrete foundation can not be accepted.

机器在基础内的灌浆是安装的一个非常重要的组成部分。不能接受灌浆复合物内的裂缝或者混凝土基础上的不良附着物。

7. Final installation and inspection 最终安装和检查

After the concrete has set, lift the machine from the foundation and retighten the anchor bolts. Check the alignment in order to ensure that the machine will run with the permissible vibration. If necessary, make the adjustment with shims, and then complete the doweling according to the holes in the feet.

在混凝土凝固后,从基础处吊装该机器并且重新拧紧地脚螺栓。检查是否找正,以确保机器能够在许可的振动范围内运行。如有必要,首先用垫片调整,然后按照地脚孔完成。

8. Testing run and start-up coordination 调试和试车

(1) All process piping and equipment related to pre-commissioning test run should be ready for operation.

与试运行有关的工艺管道及设备具备使用条件。

(2) Sharing work of electrical, steam and instrument system related to pre-commissioning test run should be compliant with the operation condition.

与试运行有关的电、汽等共用工程及仪表系统满足使用条件。

(3) Confirm that filling of lubricating oil has been finished. Inspect coupling by hand rotating. If there is no abnormal condition, inch motor to inspect rotating direction.

确认润滑油加注完毕后,应用手盘动联轴器检查,无异常情况后微动电机检查转向。

(4) Start up motor, run it according to required time and inspect the noise, bearing temperature, vibration, voltage, current, fluctuation of outlet pressure and seal. If there are abnormal conditions, immediately stop running for inspection and treatment.

启动电机,按要求时间运行并检查有无噪音,检测轴承温度、振动、电压、电流、出口压力波动情况,检查轴封等;遇异常情况立即停车检查和处理。

For different kinds of equipments, inspect every valve within test run flow when starting up pumps. Flow condition should meet the process requirement of starting different equipments. The installation testing and inspection before commissioning on site are shown in Figure 8.7 and Figure

8.8，respectively.

不同类型的设备在启动时应严格检查试运流程内各个阀门的状态。流程状态应符合不同设备启动时的工艺要求。现场的安装调试和试车前检查分别见图 8.7 和图 8.8。

Figure 8.7　Testing of installation
图 8.7　安装调试

9. Normal start-up procedure　正常启动程序

After completion of preparing work, ready for start-up (Figure 8.9).
准备工作完成后，开始启动设备（图 8.9）。

The start-up shall proceed in several steps described below. 按以下所述几个步骤启动设备。

◆ Open insert valve, close outside valve. 开入口阀，关出口阀。

◆ Start the equipment. 启动设备。

◆ check the temperature of bearing and vibration. 检查轴承温度和振动。

Figure 8.8 Inspection before commissioning

图 8.8 试车前检查

10. Equipment maintenance 设备维护

(1) Disassembling for inspection 拆检

Disassemble the equipment required for inspection (Figure 8.10). 对需要检查的设备进行解体检查（图 8.10）。

(2) Bearing inspection 轴承检查

Check damage situation of bearing surface. 检查轴承表面损伤情况。

(3) lubrication inspection 润滑检查

Check the oil color and viscosity. Check whether the oil level is less than the requirement range. 检查润滑油颜色和黏度。油位是否少于要求范围。

(4) Centering of Coupler 联轴节对中

Scene Eight Installing, Testing, Starting-up, Disassembling and Maintaining with Foreigners 情景八 与外方人员安装、调试、开车、拆检和维护

Figure 8.9 Checking start-up & run
图 8.9 启动运转检查

Figure 8.10 Imported equipment disassembly for inspection
图 8.10 进口设备拆检

◆ Laser alignment method will be adopted. Allowable deviation: radial concentricity 0.03mm/m; end parallelism 0.05mm/m. Axial distance: 2~3mm (calculating according to size of couplers). And plumb line and micrometer method will be used also.

采用激光对中方法,允许偏差:径向同心度0.03毫米/米;端面平行度0.05毫米/米。轴向间距:一般2~3mm(根据联轴节大小计算确定)。也可以采用铅垂线和千分尺等方法。

Part II Key Words and Expressions
第二部分　关键词汇

trial	试车	test run	测试运行
mechanical completion	机械竣工	operation manual	操作手册
start-up	开车/启动	nitrogen	氮气
process gas	工艺气体	rectisol	低温甲醇洗
unusual noise	异常噪音	vibration	振动
rotation number	转数	check list	检验单
quality specification	质量说明书	inspector	检查员
checker	审核人	controller	控制员
dial gauges	千分表	disassemble	解开

Part III Typical Sentences
第三部分　典型单句

1	The mechanical completion check list of the unit has been audited and approved by the representative from the buyer and seller.	这个装置的机械竣工检验表已由买方和卖方的代表审定。

Scene Eight Installing, Testing, Starting-up, Disassembling and Maintaining with Foreigners 情景八 与外方人员安装、调试、开车、拆检和维护

续表

2	We have planned to finish the adjustment of the machine before Tuesday.	我们已经计划在星期二以前完成机器的调试工作。
3	Shall we begin the test run now?	我们立即开始试车好吗？
4	something is wrong, could you get there and have a chek, please?	情况不太好，你能去检查一下吗？
5	Some parts of the machine went hot.	这机器零件发热。
6	I felt the machine shake seriously.	我感到这机器震动严重。
7	It is in good condition.	情况良好。
8	We are vevg pleased about the testing result.	试车结果使我们很满意。
9	It is not doubt that the test run will be successful.	试车将会成功是无疑的。

Part Ⅳ Situational Dialogues
第四部分 实景开口说

Liu Jie & Chen Jun：Good morning, Mr. Brown and Mr. Huebner!
　　　　　　　　布朗先生，胡博先生，早上好。

Brown & Huebner：Good morning, gentlemen!
　　　　　　　　各位早上好！

Chen Jun：Mr. Brown, when shall we put the machine to trial (test run)?
　　　　　布朗先生，我们什么时候进行试车呢？

Brown：We need to do this after finishing the erection work. But first, please make sure that the mechanical completion check list has been approved.
　　　　这台设备安装工作完成以后就可以了。但是首先，必须确保装置的机械竣工检验表得到了批准。

Liu Jie: Yes! It has been audited and approved by the representatives from both the buyer and the seller.

是的,已由买卖双方代表核定批准了。

Chen Jun: Great. And also make sure that the installation should follow the instruction and operation manual of the machine.

太好了。还要确保根据说明书和操作手册来安装。

Brown: Definitely. So, shall we finish the adjustment before Tuesday?

那肯定了。那么我们在周二前完成调试可以吗?

Huebner: No problem. Heve are something we need notice. Firstly before initial start-up of the installation, we must check the equipment carefully. Confirm that filling of lubricating oil has been finished. Inspect coupling by hand rotating, if there is no abnormal condition, inch motor to inspect rotating direction.

没问题,我们有些事情要注意。第一步,在装置初次开车以前,我们必须仔细地检查这些设备。确认润滑油加注完毕后,应用手盘动联轴器检查,无异常情况后微动电机检查转向。

Brown: Secondly, start and run the motor according to required time and inspect the noise, bearing temperature, vibration, voltage, current, fluctuation of outlet pressure and seal. If there are abnormal conditions, immediately stop running for inspection and treatment.

第二步,启动电机,按要求时间运行并检查有无噪音,检测轴承温度、振动、电压、电流、出口压力波动情况,检查轴封等;遇异常情况立即停车检查和处理。

Liu Jie: All right, Shall we begin the test run now? Are you ready?

我们立即开始试车好吗? 准备好了吗?

Huebner: Go! 好,启动!

Brown: The equipment had been running for 4 hours before carrying a full load.

这个设备在满载前已经运转了四个小时。

Huebner: Something is wrong, could you go to have a check, please?

Scene Eight Installing, Testing, Starting-up, Disassembling and Maintaining with Foreigners 情景八 与外方人员安装、调试、开车、拆检和维护

这台机器运转不好，请你能去检查一下吗？

Liu Jie: Some parts of the machine went hot.
这机器零件发热。

Brown: It's too noisy.
这台机器噪音很大。

Chen Jun: I felt the machine shake seriously.
我感到这机器震动严重。

Brown: If there is any abnormal temperature (unusual noise, vibration), it is necessary to stop the machine and figure out the reasons.
如果产生不正常的温升（异常噪音、振动），必须停车查明原因。

Chen Jun: Dismantle it for inspection.
对设备进行解体检查。

Liu Jie: I think the problem is here. We have to fix it right now.
我想故障在这里，我们必须立即修理它。

Brown: Check the oil color and viscosity. Check whether the oil level is less than the requirement range.
检查润滑油颜色和黏度。油位是否少于要求范围。

Chen Jun: Normal!
正常！

Huebner: You must turn off the switch when anything goes wrong with the motor. We should increase the pressure of the compressor gradually and put it into full load step by step.
如果电动机有什么毛病时，你必须关掉开关。我们应该渐渐地增加压缩机的压力，并逐步投入满负荷运行。

Brown: The rotation number of the machine is increasing.
机器的转数在增加。

Huebner: It is in good condition.
这台机器运转良好。

Brown: It runs perfectly well, which has been operating of 72 hours.
这台机器运转很好，它已连续运转了七十二小时。

Liu Jie: We are very pleased about the testing result.

试车结果使我们很满意。

Huebner: It is not doubt that the test run will be successful.

　　　　试车将会成功是无疑的。

Liu Jie: The controller (inspector, checker) has already signed the check list.

　　　　检验单（质量说明书）已由控制员（检查员、审核人）签字。

Brown: Great. Can't wait for that!

　　　　非常好！我们拭目以待。

Part Ⅴ　Exercises
第五部分　练习活动

Task 1: Study the following words and match the words in left column with corresponding Chinese phrases or terminologies in the right column.

____ 1. erection	a.	粗调平
____ 2. circular pitch	b.	调试
____ 3. axial distance	c.	试车/开车
____ 4. operation manual	d.	校正平面
____ 5. testing run	e.	找正
____ 6. alignment	f.	轴向间距
____ 7. rough levelling	g.	垂直定位
____ 8. correcting plane	h.	安装
____ 9. start-up	i.	操作手册
____ 10. vertical positioning	j.	齿距
____ 11. balance of rotor	k.	位移曲线
____ 12. combined mechanism	l.	振动
____ 13. phase angle of unbalance	m.	组合机构
____ 14. displacement diagram	n.	不平衡相位
____ 15. vibration	o.	转子平衡

Scene Eight Installing, Testing, Starting-up, Disassembling and Maintaining with Foreigners 情景八 与外方人员安装、调试、开车、拆检和维护

Task 2: Make sure you know how to spell and pronounce all the words above. Then write a sentence for each word. The sentence must show you understand the meaning of the word.

_____
_____
_____
_____
_____
_____
_____

Task 3: Translate the following sentences into Chinese/English.
  1. Please adjust the lift of valve plate.
  _____

  2. Shut down immediately and check the reason for abnormal sound from the compressor.
  _____

  3. Observe the figure of the compressor intake and delivery pressure.
  _____

  4. 开车前请全面检查开车条件。
  _____

  5. 请启动压缩机。
  _____

  6. 请检查仪表是否正常。
  _____

Task 4: Role Play.
Follow the example of the conversation above. Work together as a group to create an episode on this task theme. Then use role play to practice your English communication skills on installation, testing run and start-up with foreigners.

情景模拟演练任务书

班级	
任务主题	Installation, Testing Run and Start-up with Foreigners 与外方人员安装调试维护和试车
组名、成员	
演练内容	Use the expressions about this task with some communication skills (e.g. describing) to create a dialogue, an episode etc. 运用与本任务主题有关的语言表达和交流技能(如描述)创设情景会话或情景剧。
演练目的	1. master the expressions about this task. 掌握本任务的表达方式。 2. master the communication skills in English. 掌握英语交流技能。 3. put them into application. 把所学付诸于实践应用。
小组成员承担的任务及角色	
任务完成情况及步骤	
小组任务完成综合体会(综合成员个人体会)	

情景模拟演练成果展示评分标准

考核内容	分值	得分	备注
一、设计思路表达是否切合任务主题	20		
二、内容是否积极、健康、向上	20		
三、在角色表演过程中是否投入、认真、代入感强	20		
四、编选的情景模拟要自然,符合基本的英语表达习惯,是否掌握正确运用交际用语能力	20		
五、语音语调、手势表情、即兴反应	10		
六、小组成员协作与团队精神体现	10		

评价＼得分	小组最终得分	小组成员最终得分
小组互评（包括所有成员）		
教师点评		

Attachment: Episode

附： 情景剧稿

Scene Nine 情景九

On-site Operation Training
现场操作培训

Part I Background Information
第一部分 情景背景知识

Today, both automatic and mechanic degree in the project construction are increasing highly. Because the new and advances equipment is very expensive, and the operators may not be familiar with it, the equipment manufacturer or process licensor is required to provide some on-site trainings on operation and maintenance to the operators to ensure the equipment works efficiently (as shown in Figure 9.1).

当今，工程项目建设的机械化程度、自动化程度越来越高。由于新型的、先进的设备十分昂贵，而操作工可能又不太熟悉，为了保证设备的工作效率，常要求设备制造商或工艺专利商对操作工进行操作和维修方面的现场培训（见图9.1）。

As for the training process, the technical personnel of manufacturer or process Licensor will give a general introductions including: specification, performance of equipment and further detailed explanation on the equipment parts, operation steps, and guidance of operation, maintenance, repair and trouble shooting.

关于培训过程，一般制造商或工艺专利商的技术人员先对设备作总体介绍，包括设备的规格和性能。然后，进一步讲解设备组成部分、具体的操作

Figure 9.1　Equipment operation training
图 9.1　设备操作培训

步骤，以及设备的生产操作、日常保养、检修及故障排除指导等。

No matter which country the equipment are imported from, English is a general training language, and the operation and maintenance manual are also written in English. If the mechanical engineer and the operator master some skills for understanding, speaking and reading in English, it will be greatly helpful for the efficiency of work.

不管设备进口自哪国，一般培训语言常使用英语，操作维护手册也常是用英语编写的。如果现场机械工程师甚至操作工能够听懂、会说、会读这方面的英语用语，这将很大程度上提高我们的工作效率。

Example　案例

Principles of the operation of compressor 压缩机操作原则

Filtered air (controlled by an automatically operated inlet control device) enters the compressor's first stage to be compressed. After initial compression, the air travels through an intercooler where heat and moisture are removed from the compressed air.

过滤后的空气（通过自动运行的进气控制设备控制）进入压缩机第一级

进行压缩。初次压缩之后，空气穿过中间冷却器，将压缩空气中的热量和水分除去。

After the air is cooled and dried, it enters the compressor's second stage to be further compressed. The air is then directed through the second intercooler to remove heat and moisture again. Finally, the air enters the compressor's third stage, where it is compressed to the design discharge pressure and supplied to the plant air system.

空气冷却干燥之后，进入压缩机第二级进行进一步压缩。然后，空气经过引导穿过二级冷却器，继续除去热量和水分。最后，空气进入压缩机第三级，压缩到设计排放压力并供给工厂的空气系统。

The compressor's integral lubrication system supplies oil to the compressor bearings and gears, while the water system supplies cooling water to the intercooler tube bundles and the oil cooler.

压缩机的整体润滑系统向压缩机轴承和齿轮供应润滑油，同时水系统向中间冷却器管束和润滑油冷却器供应冷却水。

The PLC control system automatically positions the inlet control device to maintain the driver load within acceptable limits. Also, the unloading valve is automatically positioned to maintain the design discharge pressure. Protective devices are implemented within the control system to automatically shut down the compressor in the event that any of the operating parameters reach an unacceptable level.

PLC控制系统自动定位进气控制设备，将驱动器负荷保持在可接受的范围之内。另外，对卸载阀进行自动定位，以维持设计排放压力。保护装置在控制系统内运行，在任何运行参数达到不可接受的水平时自动关闭压缩机。

Surge is a common characteristic of centrifugal compressors. Surge occurs when the compressor is unable to overcome the pressure of the plant air system. Surge may occur under atypical design conditions or abnormal control operations. When the compressor experiences surge, the direction of airflow reverses as the compressor is attempting to produce flow against the backpressure of the plant air system. In the event of a surge, the control system automatically detects and eliminates the surge condition by unloading the compressor.

喘振是离心压缩机的共性。当压缩机无法克服工厂空气系统的压力时，发生喘振。喘振可能由非典型设计条件或异常控制运行引起。当压缩机发生喘振时，压缩机尝试产生气流对抗工厂空气系统的背压，气流方向发生反转。如果发生喘振，控制系统自动检测并通过压缩机减负消除喘振情况。

Part II Key Words and Expressions
第二部分　关键词汇

axial	轴向的	centrifugal	离心机
air separation plant	空分装置	compressor	压缩机
journal bearing	支持轴承	labyrinth seals	迷宫式密封
axial blade rows	轴向叶栅	centrifugal stage	离心级
rotor shaft	转轴	balance piston	平衡活塞
thrust bearing	推力轴承	bearing pedestal	轴承座
suction nozzle of centrifugal section	离心段吸嘴	discharge nozzle of centrifugal section	离心段喷嘴
discharge nozzle of axial section	轴向段喷嘴	casing	外壳
stator vane positioned	定子叶片定位	suction nozzle of axial section	轴向段吸嘴
maintenance	维护	trouble shooting	故障排除
performance	性能	operation manual	操作手册
replace	更换	demonstration	演示
oil filter	油滤器	air cleaner	空气过滤器

Part III Typical Sentences
第三部分 典型单句

1	Let me give you a brief description on the compressor.	让我给您简单介绍一下这台压缩机。
2	The compressor is driven by a steam turbine.	压缩机由汽轮机驱动。
3	We can easily understand what you said.	这样就好理解了。
4	How does it work?	它是怎么运转的？
5	Everyone please pay your attention!	请大家注意！
6	Have you got the *Operation Manuals*?	你们都有操作手册了吗？
7	Let's read the manual as we talk.	我们边讲边看手册。
8	Let's see what we have to do to make the compressor work well.	我们看一下该怎么做才能使压缩机良好地运转。
9	How often do we replace them?	多久更换一次？
10	We don't have any question for the moment.	暂时没有问题了。
11	We're eager to try.	我们迫不及待想试一试了。

Part IV Situational Dialogues
第四部分 实景开口说

Brown: Hello everyone. I am Brown, and this is Mr. Huebner. Today, we are going to show you how to use this compressor.
大家好，我是布朗，这位是胡博纳。今天，我们将要指导大家怎样使用这台压缩机。

Operators: Welcome!
欢迎二位！

Brown: First of all, let me give you a brief description on the compressor. The compressor is an axial/centrifugal compressor type

AR115/06L4R1 having an axial-flow low-pressure section and a radial-flow high-pressure section. It is installed in a machine train forming part of an air separation plant. It boosts the process gas pressure in the air process, and is driven by a steam turbine.

首先，让我简单介绍一下这台压缩机。这台压缩机是具有轴流式低压段和径流式高压段的轴向式/离心式 AR115/06L4R1 型压缩机。该压缩机安装在空分装置内的机组中。压缩机在空气工艺处理中对工艺气体进行增压。该压缩机由汽轮机驱动。

Operator: Excuse me, could you please show us a general picture of the compressor as you said? That will help us understand it very well.

打扰一下，您能展示一下所说的压缩机简略图吗？这样有助于我们很好地理解它。

Brown: No problem. Huebner, please pass me the drawing of compressor (Figure 9.2).

好的。Huebner 请拿一下压缩机的图纸（图9.2）。

Huebner: OK, moment please! Here you are.

稍等！给您。

Brown: Thanks, Huebner.

谢谢。

Operator: That's great. We can easily understand what you said.

好极了，这样我们就能轻易理解您的讲解了。

Operator: How does it work?

压缩机是怎么运转的？

Brown: Huebner, please give a demonstration.

Huebner，请你演示一下。

Huebner: OK, everyone please pay your attention!

好的，请大家注意！

Air is first compressed in an axial direction inside the compressor. Pressure and temperature increase from stage to stage. Having left the last axial stage, the process gas passes through an external intercooler into the centrifugal high-pressure section of the compressor, Where the process gas is compressed to discharge pressure.

Scene Nine On-site Operation Training
情景九 现场操作培训

Figure 9.2 The drawing of compressor
图 9.2 压缩机简图

NOTE:
注:

① Journal bearing 径向轴承
② Labyrinth seals 迷宫式密封
③ Axial blade rows (low-pressure stages) 1~6 轴向叶栅（低压级）1~6
④ Labyrinth seals 迷宫式密封
⑤ Centrifugal stage (high-pressure stage) 离心级（高压级）
⑥ Rotor shaft 转轴
⑦ Balance piston 平衡活塞
⑧ Journal bearing 径向轴承
⑨ Thrust bearing 推力轴承
⑩ Bearing pedestal 轴承座
⑪ Suction nozzle of centrifugal section (high-pressure stage) 离心段（高压级）吸嘴
⑫ Discharge nozzle of centrifugal section (high-pressure stage) 离心段（高压级）喷嘴
⑬ Discharge nozzle of axial section (low-pressure stages) 轴向段（低压级）喷嘴
⑭ Casing 外壳
⑮ Stator vane positioned (low-pressure stages) 定子叶片定位（低压级）
⑯ Suction nozzle of axial section (low-pressure stages) 轴向段（低压级）吸嘴

首先在压缩机内的轴向方向对空气进行压缩，压力和温度逐级增加。工艺气体离开最后一个轴向级，经过外部中冷器进入压缩机的高压离心段。在此将工艺气体压缩至排气压力。

Operator: Got it.

懂了。

Brown: Next, Huebner is going to give you some information on the maintenance and trouble shooting for the compressor.

接下来，请Huebner讲一下有关压缩机维修和故障排除方面的知识。

Huebner: Ok. I'm going to show you how to troubleshoot and maintain the compressors. All of our compressors have passed the strict inspections and tests before sale and can be used safely. However, the performance, safety, working efficiency and service life of the compressor greatly depend on your daily handling and maintenance. Have you got the *Operation Manuals*?

好的！下面请看如何排除故障和维修压缩机。所有的压缩机虽然在出厂前通过了严格的检验和测试，可以安全使用。然而，压缩机的性能、安全、工作效率及使用寿命很大程度上取决于操作人员的日常操控习惯和维护。你们都有操作手册了吗？

Operator: Yes.

有了。

Huebner: Fine. Let's read the manual as we talk. As I said just now, regular maintenance is very important to ensure that the compressor is always operated in the best conditions. Let's see what we have to do to make the compressor work well. Hydraulic oil plays a very important role in the compressor's functioning. You have to replace the oil filters and the air cleaners at regular intervals.

Scene Nine On-site Operation Training
情景九 现场操作培训

好。我们边讲边看手册。正如我刚才所说的，常规维护很重要，能够确保压缩机良好地运转。我们看一下该怎么做才能使压缩机良好运转。液压油对于压缩机的运行至关重要。必须定期更换油滤器和空气过滤器。

Operator: How often do we replace them?
多久更换一次？

Huebner: For the oil filters, 1000 working hours for the first time on a new compressor and after that, every 2000 hours. For the air cleaners, generally every 1500 hours, mainly depends on the air quality. Sometimes problems occur during the operation. Where do we have to check?

对于新压缩机而言，油滤器使用1000小时后更换，之后每2000小时更换一次。空气过滤器一般而言每1500小时更换一次，主要与空气质量有关。有时，在运转过程中发生一些问题。我们该检查哪儿呢？

Operator: I am not sure, perhaps...
不太清楚，或许……

Huebner: You can find the right answer in the trouble shooting flowchart on page 81 of the manual. And then, we go on to check whether there're any oil leakage, whether the hydraulic oil level is too low or contaminated, or whether the suction filter is clogged. So, always refer to the flowchart for help if you're not sure of how to locate the trouble.

你们可以参照操作手册第81页的故障排除流程图。我们可以检查那里是否漏油、液压油液位低或被污染，或者吸滤器被堵塞。所以，当你不确定问题所在，通常可以参照这张流程图。

Operator: How do we know it's time to overhaul the compressor?
我们如何知道何时该大修压缩机？

Huebner: Normally, only the main units were broken, like the main motor, or the compressor needs to be overhauled. The other parts of the compressor need to be checked and maintained regularly only. You can find what need to be overhauled, and how often they should be overhauled, in the manual. Any questions?

通常情况下，仅仅关键部件，比如主电机、压缩机需要大修。其他部件定期检查、维修即可。哪些部件需要大修，以及多久需要进行大修，这些都可以在手册中查到。还有什么问题？

Operator: We don't have any question for the moment.

暂时没有问题了。

Huebner: OK. Tomorrow, we will practice starting-up and shutting down the compressor.

好。明天我们演练压缩机开车和停车。

Operator: We can't wait to try, thanks a lot!

我们迫不及待想试试了。非常感谢您！

Huebner: You are welcome. See you tomorrow.

不客气。明天见！

Brown: See you, guys!

大家明天见！

Operators: See you!

再见！

Part Ⅴ Exercises

第五部分 练习活动

Task 1: Study the following words and match the words in left column with corresponding Chinese phrases or terminologies in the right column.

Scene Nine On-site Operation Training
情景九 现场操作培训

____ 1. operation training	a. 演示
____ 2. guidance of maintenance	b. 单机试运行
____ 3. instrument inspection	c. 日常保养指导
____ 4. operation steps	d. 操作培训
____ 5. operation manual	e. 仪表检查
____ 6. mechanical test run	f. 检修及故障排除指导
____ 7. guidance of operation	g. 操作步骤
____ 8. equipment operation	h. 操作手册
____ 9. demonstration	i. 生产操作指导
____ 10. guidance of repair and trouble shooting	j. 设备操作

Task 2: Make sure you know how to spell and pronounce all the words above. Then write a sentence for each word. The sentence must show you understand the meaning of the word.

_____
_____
_____
_____
_____
_____
_____
_____
_____
_____

Task 3: Translate the following sentences into Chinese/English.

1. Maintain appropriate temperature and moisture when testing.

_____

2. Strictly observe the rules of walk-around inspection and ensure the normal operation of production.

_____

3. Strength the equipment maintenance and reduce all kinds of leakages.

_____

4. 这里有异样的声音，请马上检查一下。

5. 按章操作，避免事故发生。

6. 请戴好安全帽，避免头发缠入转轮。

Task 4: Role Play.

Follow the example of the conversation above. Work together as a group to create an episode on this task theme. Then use role play to practice your English communication skills on on-site operation training.

情景模拟演练任务书

班　　级	
任务主题	On-site Operation Training 现场操作培训
组名、成员	
演练内容	Use the expressions about this task with some communication skills(e. g. describing) to create a dialogue, an episode etc. 适用与本任务主题有关的语言表达和交流技能(如描述)创设情景会话或情景剧等。
演练目的	1. master the expressions about this task. 掌握本任务的表达方式。 2. master the communication skills in English. 掌握英语交流技能。 3. put them into application. 把所学付诸于实践应用。
小组成员承担的 任务及角色	
任务完成情况 及步骤	
小组任务完成 综合体会(综合 成员个人体会)	

情景模拟演练成果展示评分标准

考核内容	分值	得分	备注
一、设计思路表达是否切合任务主题	20		
二、内容是否积极、健康、向上	20		
三、在角色表演过程中是否投入、认真，代入感强	20		
四、编选的情景模拟要自然，符合基本的英语表达习惯，是否掌握正确运用交际用语的能力	20		
五、语音语调、手势表情、即兴反应	10		
六、小组成员协作与团队精神体现	10		
评价\得分	小组最终得分	小组成员最终得分	
小组互评（包括所有成员）			
教师点评			

Attachment: Episode

附：情景剧稿

Scene Ten 情景十

Co-operation Celebration
合作庆祝

Part I Background Information
第一部分　情景背景知识

　　Generally speaking, most of the foreigners love Chinese food. Even in a pressing schedule for project constructions, we can still take advantages of the weekends or holiday to invite foreign staff to enjoy Chinese food, which can not only release their body and mind, but also express thanks for their co-operation, and enhance the working relationships as well.

　　一般来说，大多数外国人也比较喜欢中餐。即使在工期比较紧的情况下，我们也可以利用周末或节日的休息机会邀请外方工作人员，一起品尝中餐，使其放松身心，感谢与其合作，以便进一步拉近工作团队成员的合作关系。

Part II Key Words and Expressions
第二部分　关键词汇

hospitality	（热情）款待	friendship	友谊
business meal	商务宴请	banquet	宴会

Scene Ten Co-operation Celebration
情景十 合作庆祝

续表

party	聚会	tea break	茶歇
dry red wine	干红	toast	干杯
Chinese liquor	白酒	yellow rice wine or Shaoxing wine	黄酒
roasted lamb back	烤羊背	sliced mutton	冷切羊肉
noodles with grass seeds	蒿子面	deep-fried dough stick	油条
sprouts of goji	枸杞苗	purple potato	紫土豆
stewed mutton	炖羊肉	steamed batter with sauce	酿皮
beef	牛肉	mutton	羊肉
chicken	鸡肉	pork	猪肉
seafood	海鲜	duck	鸭肉
soybean milk	豆浆	barbecue	烧烤
tofu curd	豆腐脑	salted duck egg	咸鸭蛋
egg cakes	蛋饼	pancake	煎饼
syrup of plum	酸梅汤	fish filets in hot chili oil	水煮鱼
donkey burger	驴肉火烧	hot pot	火锅
dumpling	饺子	malatang or spicy hot pot	麻辣烫
wonton	馄饨	hand-pulled noodles or lamian	拉面
steamed bread/bun	馒头	steamed stuffed bun	包子
moon cake	月饼	traditional Chinese rice-pudding or zongzi	粽子
duck blood cake in chili sauce	毛血旺	cooked chopped entrails of sheep	羊杂碎
stinky tofu	臭豆腐	noodles with soybean paste	炸酱面
Chinese hamburger	肉夹馍	Beijing roast duck	北京烤鸭
sliced noodles	刀削面	tomato and egg soup	西红柿鸡蛋汤
seaweed soup	紫菜汤	hot & sour soup	酸辣汤
porridge	粥	tea	茶
mongolia milk tea	奶茶	green tea	绿茶
black tea	红茶	eight treasures tea	八宝茶

Part III　Typical Sentences
第三部分　典型单句

1	Have a seat, please.	请坐。
2	Thanks for your hospitality.	谢谢你们的热情安排。
3	Please try some of this mutton.	请品尝一下羊肉。
4	Make yourself at home.	请别拘束。
5	Very tasty!	太好吃了!
6	Help yourselves, please!	请慢用!
7	Let's toast to our health and friendship!	为我们的健康和友谊干杯!
8	I hope you have a pleasant work here.	希望你们在这里工作愉快。
9	May the friendship between us continue to grow, to the friendly co-operation and the success of the project!	为我们之间的友谊天长地久——为友好合作和项目成功干杯!
10	Do you want more?	还需要吗?
11	That's all for today, good night to all of you.	那就到这,祝大家晚安。

Part IV　Situational Dialogues
第四部分　实景开口说

Brown: Mr. Liu, thank you for your inviting.
　　　　刘先生,感谢您邀请我们。

Scene Ten Co-operation Celebration
情景十 合作庆祝

Liu Jie: It's my pleasure. Have a seat, please.
　　　　是我的荣幸。请坐。

Brown & Huebner: Thanks!
　　　　　　　　谢谢！

Liu Jie: You are welcome. Today, we are going to have some traditional Chinese food.
　　　　不客气。今天，我们为大家准备了一些特色中国美食。

Huebner: Oh, that's great.
　　　　哦，太好了！

Liu Jie: In Ningxia, the Yanchi Mutton is very famous and popular, especailly for serving guests.
　　　　在宁夏当地，盐池羊肉非常有名，特别受欢迎，尤其是招待客人必不可少。

Chen Jun: We've also prepared a bottle of Dry Red Wine for you.
　　　　　我们还为你们准备了一瓶干红。

Brown: Thank you very much for your warm heart and this wonderful dinner.
　　　 非常感谢你们的热诚和丰盛的晚餐。

Liu Jie: Our pleasure, please try some of this mutton.
　　　　非常荣幸，请品尝一下羊肉。

Brown: ... Great!... I like Chinese food very much.
　　　 哇……，好吃极了……，我非常喜欢中餐。

Chen Jun: Make yourself at home.
　　　　　请别拘束。

Huebner: Everything looks so nice. Could you tell us about these dishes, please?
　　　　 每一样菜都看起来这么鲜美。您能给我们介绍一下吗？

Chen Jun: Sure, this is Eight Treasures Tea, Roasted Lamb Back, Sliced Mutton, Noodles with Grass Seeds, Sprouts of Goji, Purple potato, Stewed Mutton, Steamed Batter with Sauce... . Help yourselves, please!
　　　　　是的，这是八宝茶、烤羊背、冷切羊肉、蒿子面、枸杞苗、紫土豆、炖羊肉、酿皮……请慢用！

Huebner: Very tasty!
太好吃了！

Liu Jie: Here comes the fish. Please have a try.
鱼上来了。请品尝一下。

Brown: What a big fish! This fish is really delicious.
好大的鱼！味道真棒。

Liu Jie: The fish is from Shahu Lake.
这是来自沙湖的鱼。

Huebner: Wow, looks attractive.
哇，看起来很诱人。

Chen Jun: I hope you have a pleasant time here.
希望你们在这里过得愉快。

Brown: Yes, we do enjoy working with you.
当然，我们很喜欢和你们一起工作。

Huebner: May our friendship last forever, to a friendly and successful cooperation.
希望我们的友谊天长地久，为友情和项目成功干杯！

Everyone: Cheers!
干杯！

Brown: We had a wonderful time tonight.
我们今晚真的很开心。

Huebner: And especially enjoy these delicious Ningxia food. Thank you again.
特别喜欢宁夏美食，再次感谢你们的招待。

Liu Jie: You are welcome. Have a good night.
不客气。晚安！

Brown: Good night! See you tomorrow.
晚安，明天见！

Scene Ten Co-operation Celebration
情景十 合作庆祝

Part V Exercises
第五部分 练习活动

Task 1: Study the following words and match the words in left column with corresponding Chinese phrases or terminologies in the right column.

____ 1. Chinese food	a. 品尝
____ 2. make yourself at your home	b. 招待客人
____ 3. business meal	c. 友好合作
____ 4. taste	d. 宴会
____ 5. help yourselves, please	e. 中餐
____ 6. entertaining guests	f. 为……干杯
____ 7. hospitality	g. 请慢用
____ 8. toast to	h. 商务宴请
____ 9. banquet	i. 热情款待
____ 10. friendly co-operation	j. 宾至如归

Task 2: Make sure you know how to spell and pronounce all the words above. Then write a sentence for each word. The sentence must show you understand the meaning of the word.

_____
_____
_____
_____
_____
_____
_____

Task 3: Translate the following sentences into Chinese/English.

1. I am very glad you like it, do have more.
   _____

2. This is a pair of Chinese chopsticks, try it and see if you like it.
   _____

3. Hope you will enjoy your stay here.
   _____

4. 为增进我们之间的友谊,为友谊和合作干杯!
   _____

5. 感谢你为促进我公司的发展而做的贡献。
   _____

6. 盐池羊肉是当地特色,特别受欢迎,尤其是招待客人必不可少的美食。
   _____

Task 4: Role Play.

Follow the example of the conversation above. Work together as a group to create an episode on this task theme. Then use role play to practice your English communication skills on celebration to co-operation.

情景模拟演练任务书

班　　级	
任务主题	Co-operation Celebration 合作庆祝
组名、成员	
演练内容	Use the expressions about this task with some communication skills (e. g. describing) to create a dialogue, an episode etc. 运用与本任务主题有关的语言表达和交流技能(如描述)创设情景会话或情景剧等。
演练目的	1. master the expressions about this task. 掌握本任务的表达方式。 2. master the communication skills in English. 掌握英语交流技能。 3. put them into application. 把所学付诸于实践应用。
小组成员承担的 任务及角色	
任务完成情况 及步骤	
小组任务完成 综合体会(综合 成员个人体会)	

Scene Ten Co-operation Celebration
情景十　合作庆祝

情景模拟演练成果展示评分标准

考核内容	分值	得分	备注
一、设计思路表达是否切合任务主题	20		
二、内容是否积极、健康、向上	20		
三、在角色表演过程中是否投入、认真、代入感强	20		
四、编选的情景模拟要自然，符合基本的英语表达习惯，是否掌握正确运用交际用语的能力	20		
五、语音语调、手势表情、即兴反应	10		
六、小组成员协作与团队精神体现	10		

得分评价	小组最终得分	小组成员最终得分		
小组互评（包括所有成员）				
教师点评				

Attachment: Episode

附：情景剧稿

Knowledge Storage

for Self-Study

自主学习

知识储备

Common Expressions and Vocabularies of Typical Imported Equipment for Energy Engineering
能源工程典型进口设备常用表达及词汇

Part I Typical communication language for Daily work
第一部分 典型岗位日常工作交流用语

I. 工艺 Process

1. 记录轴承温度。Record the bear temperature.

2. 观察压缩机进出压力的数据。Observe the figure of the compressor intake and delivery pressure.

3. 控制主冷器/蒸发器的液位。Control the main condenser/evaporator level.

4. 控制精馏塔回流比，保持流量稳定。Control reflux rate of rectification column to keep the flow rate stable.

5. 调节导叶的开度。Adjust the guide glade.

6. 您有什么建议吗？Could you give some advices?

7. 压缩机有异声，请马上停车检查。Shut down immediately and check the reason for abnormal sound from the compressor.

8. 乙烯浓度偏高。The concentration of ethylene is on the high side.

9. AI1716 的联锁值是多少？How much is the interlocking number

of AI1716?

10. 二氧化碳含量偏高，请调整工艺。The content of the carbon dioxide is too high, please adjust the process.

11. 把水泵从 A 泵切换到 B 泵。Switch the water pump from A to B.

12. 从 DCS 中观察工艺变化趋势。Observe the process changing tendency from DCS.

13. 开车前请全面检查开车条件。Please check the over-all start-up condition before start-up.

14. 请换仪表记录纸。Please change the instrument recording chart.

15. 按巡回路线进行检查。Make a walk-around inspection.

16. 请分析室派人取样测定。Ask somebody from analytical room for sampling inspection.

17. 请打开水洗系统。Please start the water washing system.

18. 氧气的熔点是多少？What is the oxygen melting point?

19. 这是粗氩系统的流程图。It is a flow sheet of the crude argon.

20. 控制油温在 42℃ 和 43℃ 之间。Control the temperature of oil between 42℃ and 43℃.

Ⅱ. 设备操作 Equipment Operation

1. 请启动压缩机。Start the compressor, please.

2. 请关闭冷凝器。Shut down the condenser, please.

3. 请启动液氧泵。Start the liquid oxygen pump, please.

4. 请调节阀门开度。Please adjust the lift of valve plate.

5. 请检查冷凝器运行情况。Inspect the operation condition of the condenser.

6. 检查压缩机轴承产生震动的原因。Check the reason for the vibration of the bearing of the compressor.

7. 请检查阀门是否良好。Please check whether the valve is in good condition.

8. 请紧固压盖螺栓。Please fasten the cap bolt.

9. 这台泵的机械密封泄露严重。The mechanical sealing of this pump leaked severely.

Common Expressions and Vocabularies of Typical Imported Equipment for Energy Engineering 能源工程典型进口设备常用表达及词汇

10. 解除联锁。Release the interlocking device.

11. 单机试运行。Mechanical test run.

12. 这台电机电流过大。The electricity of this motor is too high.

13. 请检查安全阀的泄放压力是否正常。Please check whether the relief pressure of safety valve is in good condition.

14. 立即接临时管道。Connect a temporary line at once.

15. 请检查仪表是否正常。Please check whether the instruments is in good condition.

Ⅲ. 检测 Inspection

1. 粗氮的含氧量是多少？What is the content of the crude nitrogen.

2. 对成品进行项目分析。Analyze the products with all items.

3. 用折射仪测量一下氧气的浓度。Check oxygen concentration by a refractometer.

4. 工艺条件改变后，要及时分析成品质量。Check product quality as soon as process condition changes.

5. 取样前应该用样品液清洗瓶子。Purge the bottle with samples solution before taking samples.

6. 测试时要保持一定的温度湿度。Maintain appropriate temperature and moisture when testing.

7. 请保存样品分析单一年。Keep the sample analysis record for one year, please.

Ⅳ. 安全 safety

1. 高高兴兴上班，平平安安回家。Go to work happily and go back home safely.

2. 厂区严禁吸烟。Smoking is prohibited within the plant area.

3. 进入生产现场，请戴好安全帽。Please wear safety helmet when entering into production site.

4. 不准堵塞应急通道。Don't block the emergency passage.

5. 请戴好安全帽，避免头发缠入转轮。Please wear helmet to prevent you hair round the revolver.

6. 入罐前请检查好罐内残留气体和氮气的浓度。Please check the concentration of residual gas and nitrogen before entering the tank.

7. 灭火器放在指定的地点。Fire extinguishers are kept at designated place.

8. 火警电话是119。Fire alarm calling is 119.

9. 进入控制室要有专业人员监护 It must be under supervision by professional personnel when entering control room.

10. 按章操作，避免事故发生。Operate according to the regulations to avoid accidents.

11. 有意外发生时，请立即通知调度。Please intimate dispatch department in case of an emergency.

12. 这里有异样的声音，请马上检查一下。There is a peculiar noise, check immediately.

13. 班组要定期举行安全活动。Safety activity must be organized regularly by shifts and teams.

14. 加强安全检查，防止事故发生。Strengthen routine inspections to avoid accidents.

15. 戴好防护镜，防止物料飞溅。Wear goggles to avoid splashing.

Ⅴ. 管理 Management

1. 接班前，应该先检查现场。Inspect the worksite before taking over your shift.

2. 请真实、及时的作好交接班记录。Please take notes of shift-overlap truly and timely.

3. 按时作好月报表。Make monthly report forms on time.

4. 加强岗位能力培训，提高操作能力。Strengthen on-the-job training and operation skills.

5. 严格执行巡检制度，保证生产正常运行。Strictly observe the rules of walk-around inspection and ensure the normal operation of production.

6. 减少非计划停车。Reduce the unscheduled shut-down.

7. 加强设备维护避免泄漏。Strength the equipment maintenance and reduce all kinds of leakages.

8. 各类统计数据规定保存三年。All kinds of statistical figures are kept for three years by rule.

9. 请随手关水龙头。Please turn off the tap before you leaving.

10. 减少功率的损耗。Reduce the loss of power.

11. 工完、料尽、场地清。The work is finished, materials are used up and worksite is cleared.

12. 质量第一，用户至上。Quality first, customer first.

13. 加强班组核算，提高经济效益。Strengthen the abilities of business accounting at shifts to increase economic efficiency.

14. 我们要新建一套空分装置。We are going to build a new set of ASP.

15. 上海石油化工有限公司。Shanghai Petrochemical Company Ltd.

Ⅵ. 部门 Department

1. 市场经营部 Marketing Department
2. 人力资源部 Human Resource Department
3. 财务部 Finance Department
4. 技术开发部 Technology Development Department
5. 生产部 Production Department
6. 安全监督部 Safety Supervision Department
7. 环保部 Environment Protection Department
8. 研发部 Research and Development Department
9. 炼化部 Refining and Chemical Division
10. 化工部 Chemical Division

Part II Equipment, Parts & Related Fields' Terms
第二部分 设备、部件及相关学科领域专业术语

能源工程常用设备 Energy Engineering Typical Equipment	
空分装置　ASU (air separation unit)	空压机　MAC (main air compressor)
汽轮机　ST (steam turbine)	增压机　BAC (booster air compressor)
级间冷却器　intercooler	空冷塔　air cooling tower
消音塔　silencing tower/building	空压站　air compression station
压缩机　compressor	冷冻机　refrigerator
水泵　water pump	膨胀机　expander
冷箱　cooling box	塔　tower; column
换热器　heat exchanger	洗涤塔　scrubbing tower
吸收塔　absorber	冷却塔　cooling tower
精馏塔　rectification column	蒸馏塔　distillation tower
再生塔　regeneration column	造粒塔　prflling tower
汽提塔　stripper	脱气器　degasifier
合成塔　synthesis tower	反应器　reactor
聚合釜　polymerizer	转化器、变换器　converter
脱硫反应器　desulphurization reactor	甲烷转化器　methanator
气柜　gas—holder	螺旋式气柜　helical gas—holder
湿式气柜　wet gas—holder	干式气柜　dry gas—holder
槽罐　tank	贮罐　storage tank
缓冲罐　surge tank	球罐　spherical tank
计量槽　measuring tank	接受槽　receiver tank
排污罐　blow down tank	加料槽　feed tank
汽包　steam drum	闪蒸罐　flash drum
油污罐　slop tank	溶液贮槽　solution storage tank

续表

旋风分离器 cyclone separator	分离器 separator
除氧器 deaerator	分子筛 molecular sieve
干燥器 dryer	搅拌器 agitator
萃取器 extractor	混合器 mixer
重力沉降器 gravity settler	结晶器 crystallizer
气化器 vaporizer	净化器 purifier
融解槽 melter	吸附器 adsorber
升压器 booster	去过热器 desuperheater
喷雾器 sprayer	喷射器 ejector
消声器 silencer	火炬 flare
加热器 heater	冷却器 cooler
空冷器 air cooler	水冷却器 water cooler
冷凝器 condenser	螺旋板式换热器 spiral plate heat exchanger
蒸发器 evaporator	再沸器 reboiler
急冷器 quencher	深冷器 chiller
预热器 preheater	再热器 reheater
加热器 heater	电加热器 electric heater
过热器 super heater	给水加热器 feed water heater
中间冷却器 intercooler	后冷却器 after-cooler

进口设备询价、报价及采购 Inquiry, Quotation and Procurement for Imported Equipment	
询价 inquiry	厂商报价 vendor quotation
报价书 quotation	估价 estimated price
估算 estimate	报价 quoted price
投标 bid; tender	评估 evaluation
采购 procurement; purchase	管道询价单 piping requisition sheet
订货单、订购单 purchasing order	采购说明 purchase specification
采购说明汇总表 purchasing specification summary sheet (PSSS)	请购 requisition
装箱单 packing list	交货单 delivery order(D/O)
备品备件 spare parts	预制的 prefabricated
制造者；制造厂 manufacturer; vendor	供应者 supplier
承包商 contractor	顾客 client; customer

续表

业主	owner	分包商	subcontractor
包装	packing	用户	user
防潮湿包装	moisture-proof packing	防锈包装	rust-proof packing
板条箱	crate	防水包装	water-proof packing
超尺寸运输	over-dimension cargo(ODC)	木箱	wooden box
检验	inspection	催货	expediting
船上交货,离岸价格	free on board(FOB)	运输	transportation
码头交货	ex wharf; expier	卡车交货价	free on truck
编位号的	itemized	代码	code number
短代码	short code	数量	quantity（QTY）
重量	weight	净重	net weight
毛重	gross weight	汇总表	summary sheet
工业炉及锅炉 Industrial Furnace and Boiler			
炉子	furnace	烧嘴	burner
锅炉	boiler	烟囱	stack
废热锅炉	waste heat boiler	辅助锅炉	auxiliary boiler
节能器	economizer	回热炉	direct-fired heater
辐射段	radiant section	对流段	convection section
裂化炉	reformer	焚烧炉	incinerator
回转窑	rotary kiln	炉管	furnace tube
吹灰器	soot blower	观察孔	observation door; peep door
防爆门	explosion door		
油系统 Oil System			
油站	oil station	充氮	fill nitrogen
油箱	oil tank	高位油箱	overhead tank
油冷器	oil cooler	油加热器	oil heater
蓄能器	accumulator	控制油泵	control oil pump
润滑油泵	lube oil pump	顶轴油泵	jacking oil pump
油样	oil sample	滤网	mesh
回油管线	return line	上油/进油管线	supply line
旁路/旁通	bypass	安全阀	safety valve
控制阀	control valve	槽形垫片	grooved gasket

续表

平面垫片　plane gasket	滤芯　filter element
空压机　MAC (Main Air Compressor)	
入口管道　inlet; suction pipe	滤室/滤房　filter house
排气侧　discharge side	膨胀节　compensator; bellow; expansion joint
连杆　tie rod	止逆阀　check valve
放空阀　blow-off valve (BOV)	阀体　valve body
阀杆　spindle	阀瓣　flap
人孔　manhole	喷淋管　water spray pipe
法兰对中　flange alignment	应力　stress
对不齐　mismatch; misalignment	无应力连接　stressless connection
间隙　clearance; gap	可变入口导叶　variable inner guide vane (VGV)
执行机构　actuator	叶片　blade
干扰　interference	
汽轮机　ST (Steam Turbine)	
联轴器/联轴节[①]　coupling	联轴器罩　coupling guard
联轴器轮毂　coupling hub	联轴器对中　coupling alignment
激光对中仪　laser alignment device	盘车装置/盘车电机　turning gear (RTG: rotating turning gear)
轴承　bearing	滑动轴承/径向轴承　journal bearing
轴承箱　bearing house	转子　rotor
推力轴承/止推轴承　thrust bearing	入口侧/进气侧　suction side
出口侧/排气侧　discharge side	缸体　casing
吹扫假件　blow-off device/Blowing device	速关阀　quick-close valve (QCV)
保温层　insulation	电液转换仪　electric-hydraulic converter (EH Converter)
调速阀　control valve	疏水器/疏水阀　trap
导淋/疏水　drain	末级导叶　last blade
乏汽管道　exhaust duct	就地控制面板　local control panel (LCP)
次末级导叶　secondary last blade	冷态对中　cold alignment
轴承座　pedestal	喷油嘴　oil spray nozzle
增压机　BAC (Booster Air Compressor)	
冷却器　cooler	级间/中间冷却器　intercooler
级间管道　inter-stage pipe	配管　piping

续表

蜗壳　volute	入口导叶/进气导叶　inlet guide vane (IGV)
叶轮　impeller	小齿轮　pinion
扩压器　diffuser	溅油环　splash ring
蒸汽系统　Steam System	
高压蒸汽管道支架　HP steam pipe supports	弹簧吊架　spring hanger
弹簧支架　spring support	恒力弹簧支架　constant spring support
限位器　stopper	冷态　cold state/condition
热态　hot state/condition	位移/位移量　displacement
管夹/管卡　pipe clamp	间距　gap
高压密封蒸汽　HP seal steam	低压密封蒸汽　LP seal steam
轴封蒸汽　gland steam	启动抽气器　start-up ejector
运行抽气器/主抽　operation ejector	冷凝系统　condensing system
真空破坏器　vacuum breaker	冷凝水　condensate
流向　flow direction	抽真空　vacuum-pumping
阳离子　cation	阴离子　anion
氢电导率　cation conductivity	直接电导率　direct conductivity
氟化物　fluoride	磷酸根　phosphate radical
氯化物　chloride	硫酸盐　sulfate
硅酸盐　silicate	消音器　silencer
消间塔　silencing tower/building	乏气　exhaust
汽水分离器/旋风分离器　cyclone separator	梯度　gradient
零部件　Parts and Components	
螺栓　bolt	螺母　nut
丝堵/盲塞　blind plug	接头　connector
垫片　gasket	石墨垫片　graphite gasket
金属缠绕垫片　spiral wound gasket	齿形垫片　serrated gasket
石棉垫片　asbestos gasket	垫圈　washer
连杆　tie rod	电缆扎带　cable tie
世伟洛克　swagelok[②]	孔板　orifice
平台　platform	测压接头　measurement coupling
软管　hose; flexible hose	格兰头　cable gland 电缆进接线箱的部分，用于固定和保护电缆

Common Expressions and Vocabularies of Typical Imported Equipment for Energy Engineering 能源工程典型进口设备常用表达及词汇

续表

细管	tube	三通	tee(connection)
对丝	double nipple	丝扣	screw thread
机械安装/调试 Mechanical Erection/Commissioning			
检网	check the mesh	取油样	take oil samples
油分析	oil analysis	冷却油泵	cooling oil pump
泵切换	pump switchover	电机保护时间	motor protection time
上电	power on	断电	power off
启动	start	启动/开机/开车	start up
开	start; open	关	close; Shutdown
紧停	emergency shutdown(ESD)	紧停按钮	ESD button
单试/冲转	solo run	停车	shutdown
转速	speed; rpm(revolution per minute)	提速	ramp up
升压/降压	Pressurize/depressurize	冷却	cool down
暖管	pipe warming up	温度	temperature
阀腔	valve chest	主蒸汽隔离阀	main isolation valve
振动	vibration	液态	liquid state
流量	flux	蒸发	evaporation
冷凝	condensation	精馏	rectification
蒸馏	distillation	自动的	automatic
手册	manual	加热	heating
冷吹	cooling	隔离	isolation
卸压	depressurize	泄压	release the pressure; pressure release
漏点	leakage	大齿轮	gear wheel
齿轮箱	gear box	激光焊接	laser welding
总阀	main valve	X射线探伤	X-ray detection
跳车	trip	气穴	cavitation
着色探伤	dye penetrant inspection	压降	pressure drop
停转时间/惰转时间	run down time	大气安全阀	atmospheric relief valve
压力控制阀	pressure control valve (PCV)	轴承	bearing
压力	pressure	超速试验	overspeed test
高	high	防喘振试验	anti-surge test
低	low	油脂	grease

续表

临界转速	critical speed	内窥镜	endoscope/borescope
管束	bundle	工艺管道	process pipe
压力表	pressure gauge	公差	tolerance
排气	venting	非啮合	disengaged
啮合	engaged	齿轮间隙	tooth clearance/gap
齿对齿	tooth to tooth	滑动点	sliding point
单试盘	solo plate	盲板	blind plate
固定点	fixing point	槽钢	channel steel
角钢	angle steel	靶板	target plate
闪蒸罐	flash tank	通径	DN
空冷器	Air Cooling Condenser	预拉	prestretching
工厂验收试验	factory acceptance test (FAT)	填料函	stuffing box
氩焊	argon welding	珠光砂	perlite
平行度	parallelism	蒸汽	steam
水	water	氧	oxygen
油	oil	氩	argon
氮	nitrogen	乙烯	ethylene
二氧化碳	carbon dioxide	滑动鞍座	sliding saddle
固定鞍座	fixed saddle	人孔	manhole (MH)
管口	nozzle	检查孔	inspection hole
手孔	handhole (HH)	切线	tangent line
吊柱	davit	件号	part number
焊接线	weldingiine	加强圈	reinforcing ring
椭圆形封头	ellipsoidal head	吊耳	lifting lug
接地板	earth lug	轴	shaft
气缸	cylinder	附件,附属设备	accessory
轴承	bearing	无位号设备	non-itemized equipment
编位号设备	itemized equipment	阀井	valve pit
清扫	clean out	沉淀器	settler
软化装置	softener	脱盐装置	demineralizer
沉淀池	clarifier	中和池	neutralization tank
渗滤坑	soak-away pit	井	well

续表

冷却塔	cooling tower	安装	erection; installation
水处理	water treatment	试车	commissioning; running-in
小时工	man-hour (M/H)	开车	start-up
日工	man-day (M/D)	运行中	on stream
月工	man-month (M/M)	停车	shut-down
检验及试验 Inspection and Test			
焊接检验	welding inspection	无损检验	non-destructive testing (NDT)
肉眼检验；外观检验	visual inspection	着色渗透检验	dye penetrant inspection
液体渗透检验	liquid penetrant test	γ射线照相	gamma radiography
X射线照相	X-ray radiography	射线检验	radiographic testing (RT)
涡流探伤	eddy current testing (ET)	超声波探伤	ultrasonic testing (UT)
磁粉探伤	magnetic particle testing (MT)	压力试验	pressure testing
水压试验	hydraulic test	气压试验	air pressure test
气密试验	airtightness test	泄漏试验	leakage test
卤气泄漏试验	halogen gas leak test	盛水试验	full water test
真空试验	vacuum test	拉伸试验	tension test
弯曲试验	bending test	冲击试验	impact test
硬度试验	hardness test	扩口试验	flaring test
弯曲试验	bending test	拉伸试验	tension test
焊接工艺评定试验	welding procedure qualification test	冲击试验	impact test
腐蚀试验	corrosion test	硬度试验	hardness test
水压试验	hydraulic test	泄漏试验	leak test
仪表 Instrument			
技术控制中心	TCC (technology control center)	振动探头	vibration probe
温度探头	temperature probe	轴位移探头	axial displacement probe
转速探头	speed probe	接线箱/接线盒	junction box
端子	terminal	电缆	cable
间隙电压	gap voltage	电缆格兰头	cable gland
电缆扎带	cable tie	安全仪表回路	safety instrument loops (SIL)
安全仪表系统	safety instrument system (SIS)	模拟	simulation
回路测试	loop test	软启动器	soft starter

续表

前置放大器	pre-amplifier	频率发生器	frequency generator
HART 通讯器	HART communicator	振动模拟器	vibration simulator
支线	feeder	反馈	feedback
跳闸/跳车	trip	压力变送器	pressure transmitter
允许启动	release-on	逻辑与功能测试	logic and function test
允许停车	release-off	故障安全	fail-safe
保护启动	protection-on	强制	force
保护停车	protection-off	取消强制	remove the force; unforce
DCS 模式	DCS mode	循环逻辑	circular logic
TCC 模式	TCC mode	ABB 定位器	ABB positioner
限位开关③	limit switch	减压器	pressure reducer
机组控制面板	unit control panel (UCP)	电磁阀	solenoid valve
印制线路板	printed circuit board (PCB)	防喘振阀	anti-surge valve
密封气	seal air	人机界面	human-machine interface (HMI)
仪表气	instrument air	跳线	jumper
公共跳车	common trip	接线	wiring
触点	contactor	本特利系统	Bentley system
布朗系统	Braun system	超速保护系统	OVS
状态监测系统	status monitoring system	离散控制系统/集成涡轮机和压缩机控制系统	Distributed Control System/Integrated Turbine & Compressor Control System (DCS/ITCC)
中控室	central control room (CCR)	在关闭/开启状态下锁定	locked closed/locked open (LC/LO)
高速/低速	high speed/low speed (HS/LS)	入口导叶	inlet guide vane (IGV)
可编程序逻辑控制	programmable logic control	可变入口导叶	variable inlet guide vane
电机控制中心	motor control center	直行程	linear
机器监控系统	machine monitoring system	部分行程测试	partial stroke test
角行程	rotary	膜片	diaphragm
安全栅	barrier	阻尼	damping
保险丝	fuse	油阻尼器	Oil damper
光纤交换机	fiber switcher	正常供电	normal power supply (NPS)
死区	dead zone/death band	不间断电源	uninterrupted power supply (UPS)

续表

转向测试　direction test	界面　interphase
示波器　oscilloscope	压差信号　DP signal
故障开/故障关④　FO/FC, fail open/fail close	低选器⑤　minimum selector
启动升速控制器　startup ramp controller	位号　tag number
高选器　maximum selector	就地盘　local panel
盘（操作盘）　panel	常开　normally open
仪表盘　instrument panel	仪表电缆槽（架）　instrument cable tray (duct)
常闭　normally closed	压力计　manometer; pressure gauge
电缆槽（架）　cable tray (channel); cable rack	温度计　thermometer
热电偶　thermocouple	玻璃液位计　glass level gage
液位计　level gauge	流量计　flow meter
温度计保护管　thermowell	变送器　transmitter
孔板　orifice	联锁　interlock
仪器；设备　apparatus	仪器分析　instrumental analysis
自动分析　automatic analysis	气体分析　gas analysis
化学分析　chemical analysis	气相色谱仪　gas chromatograph
质谱分析　mass spectrometric analysis	膨胀端压力　expander tip pressure
液相色谱仪　liquid chromatograph	变速箱前润滑油压力　lube oil pressure before box
膨胀机出口压力　expander outlet pressure	

P&ID 图难点词汇　P&ID Key Words	
crusher　粉碎机	convey　搬运，传达，转让
ejector　驱逐者，放出器，排出器，喷淋器	centrifuge　离心分离机
compressor　压缩机	bin　箱柜
agitation　搅动	incinerator　焚化装置，焚化炉
flare　闪光，闪耀 火炬	flame　火焰
filter　过滤器	fan　鼓风机，风扇
elevator　电梯，升降机	mill　磨粉机
mixer　搅拌器，[电]混频器	sieve　筛，滤网
sewer　下水道	reducer　异径管
reactor　反应器	pelletizer　造球[粒]机，树脂料切碎机
package　包设备	orifice　孔，口

续表

motor 发动机，电动机	silencer 消音器，灭音器
slope 斜坡，斜面，倾斜	spraye 喷雾，喷雾器
strainer 滤网，松紧扣，过滤器	sump 污水坑，水坑，池，机油箱
temperature 温度	vessel 容器，器皿，导管
法兰 Flange (FLG)	
整体管法兰 integral pipe flange	钢管法兰 steel pipe flange
螺纹法兰 threaded flange	滑套法兰（包括平焊法兰） slip-on flange (SO); slip-on welding flange
承插焊法兰 socket welding flange	松套法兰 lap joint flange (LJF)
对焊法兰 welding neck flange (WNF)	法兰盖 blind flange; blind
孔板法兰 orifice flange	异径法兰 reducing flange
盘座式法兰 pad type flange	松套带颈法兰 loose hubbed flange
焊接板式法兰 welding plate flange	对焊环 welding neck collar（与 stub end 相似）
平焊环 welding-on collar	突缘短节 stub end; lap
翻边端 lapped pipe end	松套板式法兰 loose plate flange
压力级 pressure rating; pressure rating class	压力-温度等级 pressure-temperature rating
法兰密封面/法兰面 flange facing	突面 raised face (RF)
凸面 male face (MF)	凹面 female face (FMF)
榫面 tongue face	槽面 groove face
环连接面 ring joint face	全平面；满平面 flat face; full face (FF)
光滑突面 smooth raised face (SRF)	法兰面加工 facing finish
粗糙度 roughness	光滑的 smooth
齿形 serrated	均方根 root mean square (RMS)
算术平均粗糙高度 arithmetical average roughness height (AARH)	配对法兰 companion-flange
螺栓圆 bolt circle	
垫片 Gasket (GSKT)	
垫片的型式 type of gasket	平垫片 flat gasket
环形平垫片 flat ring gasket	平金属垫片 flat metal gasket
夹棉织物的橡胶 elastomer with cotton fabric insertion	夹石棉织物的橡胶 elastomer with asbestos fabric insertion
夹石棉织物及金属丝加强的橡胶 elastomer with asbestos fabric insertion and with wire reinforcement	无石墨压缩白石棉垫片 non graphited compressed white asbestos gasket

续表

天然白橡胶垫片　natural white rubber gasket	压缩石棉垫片　compressed asbestos class gasket
浸聚四氟乙烯的石棉垫片　PTFE impregnated asbestos gasket	夹石棉的缠绕金属垫片　spiral-wound metal gasket with asbestos filler
内环　inner ring	外环　outer ring
波纹金属垫片　corrugated metal gasket	波纹金属包嵌石棉垫片　corrugated metal gasket with asbestos inserted
双夹套波纹金属包石棉垫片　corrugated metal double jacketed asbestos filled gasket	双夹套垫片　double jacketed gasket
金属包石棉平垫片　flat metal jacketed asbestos filled gasket	整体金属齿形垫片　solid metal serrated gasket
槽形金属垫片　grooved metal gasket	环形连接金属垫片　ring joint metal gasket
八角环形垫片　octagonal ring gasket	椭圆环形垫片　oval ring gasket
透镜式垫片　lens gasket	非金属垫片　non-metallic gasket
阀门　**Valve**	
阀轭　yoke	外螺纹阀杆及阀轭　outside screw and yoke (OS & Y)
阀杆　stem	内螺纹　inside screw (IS)
阀轭套　yoke sleeve	阀杆环　stem ring
阀座　valve seat (body seat)	阀座环、密封圈　seat ring
整体(阀)座　integral seat	堆焊(阀)座　deposited seat
阀芯(包括密封圈、杆等内件)　valve plug	阀盘　disc
阀盘密封圈　disc seat	阀体　body
阀盖　bonnet	阀盖衬套　bonnet bush
螺纹阀帽　screw cap	螺纹阀盖　screw bonnet
螺栓连接的阀盖　bolted bonnet (BB)	活接阀盖(帽)　union bonnet (cap)
螺栓连接的阀帽　bolted cap (BC)	焊接阀盖　welded bonnet (WB)
本体阀杆密封　body stem seal	石棉安全密封　asbestos emenen seal
倒密封　back seal	压力密封的阀盖　pressure-tight bonnet
动力操纵器　powered operator	电动操纵器　electric motor operator
气动操纵器　pneumatic operator	液压操纵器　hydraulic operator
快速操纵器　quick-acting operator	滑动阀杆　sliding stem
正齿轮传动　spur gear operated	伞齿轮传动　bevel gear operated
扳手操作　wrench operated	链轮　chain wheel
手轮　hand wheel	手柄　hand lever (handle)
气缸(或液压缸)操纵的　cylinder operated	链条操纵的　chain operated
等径孔道　full bore; full port	异径孔道　reducing bore; reduced port; venturi port
短型　short pattern	紧凑型(小型)　compact type

续表

笼式环　lantern ring	压盖　gland
阀杆填料　stem packing	阀盖垫片　bonnet gasket
升杆式（明杆）　rising stem (RS)	非升杆式（暗杆）　non-rising stem (NRS)
指示器/限位器　indicator/stopper	注油器　grease injector
可更换的阀座环　renewable seat ring	

常用阀　Commonly Used Valve		
闸阀 gate valve	平行双闸板　double disc parallel seat	开口楔形闸板　split wedge
	挠性整体楔形闸扳　flexible solid wedge	整体楔形闸板　solid wedge
	塞型闸阀　plug gate valve	直通型闸阀　through conduit gate valve
截止阀 globe valve	球心型阀盘　globe type disc	塞型阀盘　plug type disc
	可转动的阀盘　swivel disc	
球阀 ball valve	三通球阀　three-way ball valve	装有底轴的　trunnion mounted
	耐火型　fire safe type	浮动球型　floating ball type
	防脱出阀杆　blowout proof stem	
蝶阀 butterfly valve	对夹式（薄片型）　wafer type	凸耳式　lug type
	偏心阀板蝶阀　offset disc burerfly valve; eccentric butterfly valve	斜阀盘蝶阀　canted disc butterfly valve
	连杆式蝶阀　link butterfly valve	
旋塞阀 plug valve	三通旋塞阀　three-way plug valve	四通旋塞阀　four-way plug valve
	旋塞　cock	衬套旋塞　sleeve cock
隔膜阀 diaphragm valve	橡胶衬里隔膜阀　rubber lined diaphragm valve	直通式隔膜阀　straight way diaphragm valve
	堰式隔膜阀　weir diaphragm valve	
止回阀 check valve	升降式止回阀　lift check valve	旋启式止回阀　swing check valve; flap check valve
	落球式止回阀　ball check valve	弹簧球式止回阀　spring ball check valve
	双板对夹式止回阀　dual plate wafer type check valve	无撞击声止回阀　non-slam cheek valve
	底阀　foot valve	切断式止回阀　stop check valve; non-return valve
	活塞式止回阀　piston check valve	斜翻盘止回阀　tilting disc check valve
	蝶式止回阀　butterfly check valve	密封　seal

续表

其他用途的阀 other valves	安全阀 safety valve (SV)	卸压阀 relief valve (RV)	
	安全卸压阀 safety relief valve	杠杆重锤式 lever and weight type	
	先导式安全阀 pilot operated safety valve	复式安全泄气阀 twin type safety valve	
	罐底排污阀 flush-bottom tank valve	电磁阀 solenoid valve; solenoid operated valve	
	电动阀 electrically operated valve; electric-motor operated valve	气动阀 pneumatic operated valve	
	低温用阀 cryogenic service valve	蒸汽疏水阀 steam trap	
	机械式疏水阀 mechanical trap	浮桶式疏水阀 open bucket trap; open top bucket trap	
	浮球式疏水阀 float trap	倒吊桶式疏水阀 inverted bucket trap	
	自由浮球式疏水阀 loose float trap	恒温式疏水阀 thermostatic trap	
	金属膨胀式蒸汽疏水阀 metal expansion steam trap	液体膨胀式蒸汽疏水阀 liquid expansion steam trap	
	双金属膨胀式蒸汽疏水阀 bi-metallic expansion steam trap	压力平衡式恒温疏水阀 balanced pressure thermostatic trap	
	热动力式疏水阀 thermodynamic trap	脉冲式蒸汽疏水阀 impulse steam trap	
	放气阀（自动放气阀） air vent valve (automatic air vent valve)	平板式滑动闸阀 slab type sliding gate valve	
	盖阀 flat valve	换向阀 diverting valve; reversing valve	
	热膨胀阀 thermo expansion valve	自动关闭阀 self-closing valve	
	自动排液阀 self-draining valve	管道盲板阀 line-blind valve	
	挤压阀 squeeze valve（用于泥浆及粉尘等）	呼吸阀 breathing valve	
	风门、挡板 damper	减压阀 pressure reducing valve; reducing valve	
	控制阀 control valve	膜式控制阀 diaphragm operated control valve	
	执行机构 actuator	背压调节阀 back pressure regulating valve	
	差压调节阀 differential pressure regulating valve	压力比例调节阀 pressure ratio regulating valve	

续表

未指明结构(或阀型)的阀 unspecified structure valve	切断阀 block valve; shut-off valve; stop valve		调节阀 regulating valve
	快开阀 quick opening valve		快闭阀 quick closing valve
	隔断阀 isolating valve		三通阀 three-way valve
	夹套阀 jacketed valve		非旋转式阀 non-rotary valve
	排污阀 blowdown valve		集液排放阀 drip valve
	排液阀 drain valve		放空阀 vent valve
	卸载阀 unloading valve		排出阀 discharge valve
	吸入阀 suction valve		多通路阀 multiport valve
	取样阀 sampling valve		手动阀 hand-operated valve; manually operated valve
	锻造阀 forged valve		铸造阀 cast valve
	(水)龙头 bibb; bib; faucet		抽出液阀(小阀) bleed valve
	旁路阀 by-pass valve		软管阀 hose valve
	混合阀 mixing valve		破真空阀 vacuum breaker
	冲洗阀 flush valve		第一道阀;根部阀 primary valve; root valve
	总管阀 header valve		事故切断阀 emergency valve
节流闪阀 throttle valve			针阀 needle valve
角阀 angle valve			Y型阀(Y型阀体截止阀) Y-valve (Y-body globe valve)
柱塞阀 piston type valve			夹紧式胶管阀 pinch valve
设备布置 Equipment Layout			
设备位号 equipment item number			工厂 plant
工厂区界 plant limit			项目区界内侧 inside battery limit
项目区界 battery limit (BL)			区界 area limit
区域边界 zone limit			界外 off site
上 up			下 down
工厂北向 plant north			真实北向 true north
道路 road			小过道 cat walk; cat way
走道、过道 walk way; gangway; access way			污水坑(井) sump pit
集水池 catch basin			沟槽 through
预留区 trough			预留场地 future area
铺砌区 paving area			非铺砌区 unpaved area

续表

碎石铺面 gravel paving	橡皮铺面 rubber paving
面积、区域 area	体积、容积 volume (VOL)
管道布置 piping Layout	
管道设计 piping design	管道研究 piping study
走向研究 routing study	重要管道 critical piping
地上管道 above ground piping	地下管道 under ground piping
管网 network of pipes	管廊 pipe rack
管沟 piping trench	管间距 line spacing
管道跨距 line span	排液 drain
旁路 by pass	上升管、垂直管 riser
放空 vent	裸管 bare line
导管 conduit	管段、接口 spool piece; spool
管件直接 fitting to fitting (FTF)	集液包 drip leg
取样接口 sampling connection	蒸汽伴热 steam tracing
伴热管 tracing pipe	电伴热 electrical tracing
热水伴热 hot-water tracing	全夹套的 full jacketed
夹套管 jacketed line; jacketed piping	详图 detail drawing
平面图 plan	剖视 A-A section "A-A"
视图 X view "X"	接续图 continue on drawing (COD)
连接图 hook up drawing	比例 scale
接续线 match line	符号 symbol
图例 legend	方位 orientation
管口表 list of nozzles	定位 location
管口方位 nozzle orientation	水平的 horizontal
相交 intersection	垂直、正交、垂直的 perpendicular
垂直的、立式的 vertical	水平安装 horizontal installation
平行、平行的 parallel	对称的 symmetric
垂直安装 vertical installation	顺时针方向 clock wise
相反（的）、对面（的） opposite	计算机辅助设计 computer aided design (CAD)
逆时针方向 counter clock wise	
图面标注 Sheet Mark	
绝对标高 absolute elevation	海平面标高 over-sea mean level (OSL)

续表

标高　elevation (EL)	混凝土顶面　top of concrete
架顶面　top of support (TOS)	钢结构顶面　top of steel
梁顶面　top of beam (TOB)	支撑点　point of supporting (POS)
管顶　top of pipe (TOP)	管底　bottom of pipe (BOP)
沟底　bottom of trench	管子内底　invert (inside bottom of pipe)
底平　flat on bottom (FOB)	顶平　flat on top (FOT)
工作点　working point	面至面　face to face (F-F)
中心至端面　center to end (C-E)	中心至面　center to face (C-F)
中心至中心　center to center (C-C)	坐标　coordinate
坐标原点　origin of coordinate	出口中心线　center line of discharge
入口中心线　center line of suction	中心线　centerline
入口　inlet; suction	出口　outlet; discharge
排气　exhaust	距离　distance

工艺　Process		
图及表 diagram and list	流程图　flow diagram	工艺流程图　process flow diagram (PFD)
	管道及仪表流程图　piping and instrument flow diagram (P&ID)	公用工程流程图　utility flow diagram (UFD)
	管线表 line list; line schedule	设备表 equipment list (schedule)
流体 fluid	空气　air	仪表空气　instrument air
	工艺空气　process air	低压蒸汽　low pressure steam
	中压蒸汽　medium pressure steam	高压蒸汽　high pressure steam
	伴热蒸汽　tracing steam	饱和蒸汽　saturated steam
	过热蒸汽　superheated steam	氧气　oxygen
	氢气　hydrogen	氮气　nitrogen
	燃料气　fuel gas	天然气　natural gas
	火炬气　flare gas	酸性气　sour gas
	液化石油气　liquefied petroleum gas (LPG)	氨气　ammonia gas
	冷却水　cooling water	循环水　circulating water
	锅炉给水　boiler feed water	热水　hot water
	蒸汽冷凝水　steam condensate	盐水　brine

Common Expressions and Vocabularies of Typical Imported Equipment for Energy Engineering 能源工程典型进口设备常用表达及词汇

续表

流体 fluid	工艺水 process water	化学污水 chemical sewage
	防腐剂 corrosion inhibitor	重油 heavy oil
	石脑油 naphtha	燃料油 fuel oil
	润滑油 lubricating oil	密封油 sealing oil
	冷冻剂 refrigerant	载热体 heating medium
	溶剂 solvent	溶液 solution
	母液 mother liquor	单体 monomer
	聚合物 polymer	均聚物 homopolymer
	共聚物 copolymer	工艺液体 process liquid
	工艺气体 process gas	硫酸 sulphuric acid
	盐酸 chlorhydric acid	硝酸 nitric acid
	烧碱 caustic soda	
流体特性 fluid characteristics	蒸汽压 vapor pressure	临界温度 critical temperature
	临界点 critical point	临界压力 critical pressure
	比热 specific heat	湿度 humidity
	密度 density	比重 specific gravity
	黏度 viscosity	闪点 flash point
	熔点 melting point	凝固点 freezing point
	浓度 concentration	爆炸极限 limit of explosion
	有毒的 toxic	可燃的、易燃的 flammable
其他 others	回收 recovery	再生 regeneration
	循环 circulation	再循环、再生 recycle
	补充 make-up	制备 preparation
	蒸汽吹扫 steaming out	涂除 purge
	抽空、排空 evacuation	吹出 blow-off
	排污 blow down	破真空 vacuum breaker
	大气腿 barometric leg	备用 stand-by; spare
	化学清洗 chemical cleaning	净正吸入压头 net positive suction head (NPSH)
	进料 feed	成品 product
	污染 contamination	大气污染 atmospheric pollution
	环境温度 ambient temperature	

续表

涡轮；透平	turbine	离析器；分离器	separator
金属小球；使变颗粒状	prill	溶液；解决方案	solution
蒸发(作用)	evaporation	循环，流通	circulation
刮料机；平土机；铲土机	scraper	再沸器	reboiler
转化炉	converter	冷凝器；电容器	condenser
交换器；换热器	exchanger	滤网，过滤器	strainer
(使)浓缩；精简	condense	发展的进程、阶段或时期	stage
冷却器	cooler	吸收塔	absorber
(盛液体、气体的大容器)罐；槽；储水池	tank	调压器；升压器	booster
[生化]分解体(分解已败死的原生质之有机体)分解器	decomposer	排出器；喷射器	ejector
泵；抽水机	pump	控制	control
蒸汽；水汽	steam	分离器	separater
量；数量	quantity	压力	pressure
技术；方法	technique	专业的	special
容器；导管	vessel	再流通、循环	recirculation
压缩机	compressor	化学制品；化学药品	chemical
量度标准；标准化	standardization	焊接工	welder
运输；输送	conveying	工业；产业；行业	industry
热；加热	heat	焊接；焊缝	weld
处理	treatment	检查；视察	inspection
材料；原料	material	阀，[英]电子管，真空管	valve
分析；分解	analysis	弹性；适应性；机动性	flexibility
轮缘；法兰	flange	大气	atmosphere
工程术语 Engineering Terms			
建筑物	building	钢结构	steel structure
已有钢结构	existing steel structure	钢筋混凝土结构	reinforced concrete construction
楼面	floor	平台	platform (PF)
地面水平	ground level	栏杆	handrail
直梯	ladder	楼梯	stair；stair way
圆柱	column	基础	foundation；footing

续表

梁 beam	构件 member
斜撑；支撑 bracing	箅子板 grate；grating
桁架；主梁 girder	吊装孔 erection opening
隔墙 partition wall	吊梁 hoisting beam
棚 shelter	窗 window
天窗 skylight	门 door
防火门 fire door	地漏 floor drain
防火层 fire-proofing	总图 general plot plan
承担 shoulder	打桩 pile
灌浆 grouting	风荷载 wind load
雪荷载 snow load	动力荷载 dynamic load
静荷载 dead load	活荷载 live load
风速 wind velocity	主导风向 most frequent wind direction
地基基础类 ground foundation	地基 foundation
地震震级 earthquake magnitude	地耐力 earth (ground) endurance
地质钻探 geological prospecting	钢筋砼基础 reinforced concrete
砖基础 brick foundation	桩基础 pile foundation
深基础 deep foundation	浅基础 shallow foundation
地梁 ground beam	基础梁 foundation beam
防潮层 damp-proof course	机械碾压地基 machine-rolled foundation
基坑边坡 slope of foundation pit	钻孔灌注桩 drilling-poured pile
爆破灌注桩 explosion-poured pile	钢板桩 steel sheet pile
壳体基础 shell foundation	脚手架 scaffold
钢脚手架 steel scaffold	扣件式钢管脚手架 fastener steel pipe scaffold
螺栓式钢管脚手架 bolt steel pipe scaffold	立杆 upright stanchion
横杆 horizontal pole；transverse strut	满堂脚手架 all-round scaffold
上料平台架 material-putting platform	活动平台架 moving platform
挑檐脚手架 cornice scaffold	吊（悬空）脚手架 hanging scaffold
折叠式里脚手架 folding internal scaffold	外脚手架 external scaffold
木脚手板 wooden scaffold board	安装图 installation diagram
吊盘 sling tray	型钢脚手板 shaped steel scaffold board

续表

门架	gantry	井架	derrick
工作台	work bench	砂垫层	sand bedding course
呈台底层;基(础)层	footing course	筏式基础	raft foundation
刚性基础	rigid foundation	素土垫座	plain concrete saddle
毛石基础	rubble foundation		
消防 Fire Fighting			
消火栓	hydrant	水炮	water monitor
洒水器	sprinkler	水喷淋	watering
消防泵	fire pump	灭火器	fire extinguisher
泡沫消防	foam fire-fighting	消防车	fire fighting truck
泡沫炮	foam monitor	泡沫栓	foam hydrant
消防软管接头	fire hose connection		
电 Electricity			
危险区划分	hazardous area classification	危险区平面图	hazardous area plan
电气盘	electrical panel	配电盘	switch board
电缆沟	cable trench	电流	current
直流电	direct current	交流电	alternating current
频率	frequency	相位	phase
电压	voltage	功率因数	power factor
电容	capacitance	电阻	resistance
接地	grounding; earthing	静电	static electricity
避雷针	lightning rod	照明	lighting; illumination
电话	telephone	电线	wire
导线管	conduit tube	接线箱(盒)	junction box
电度表	watt-hour meter	万能表	universal meter
钳形电流表	pincerlike galvanometer	安全带	safety-belt
电缆	cable	脚扣子	pole climbers
紧线器	wire grip	电线	wire
防水弯头	waterproof elbow	接线盒	wiring case
插座	socket	拉线开关	pull switch
布线	wiring	管夹	pipe clip
扳动开关	switching	安全开关	safety switch

续表

螺钉　screw	木螺钉　wood screw
电源；动力供应　power supply	变电站(所)　substation
配电盘　switch panel	油(浸)开关　oil switch
继电器　relay	钢模　die
保险丝　fuse	电路　circuit
低压　low voltage	高压　high voltage
总开关　master switch	变压器　transformer
绝缘胶带　insulating tape	功率　power
绝缘材料　insulating materials	绝缘体　insulator
额定电压　rated voltage	额定电流　rated current
额定功率　rated power	额定负载　rated load
闸刀开关　knife switch	屏蔽　shielding
屏蔽线　shielding wire	电气平面图　electric plane diagram
电气系统图　electric system drawing	电气盘面图　electric plate schema
电气原理图　electric principle drawing	电气接线图　electric wiring diagram
电气施工大详图　electrical construction detail drawing	配电盘结构图　distributor structure drawing
管子　Pipe	
管子(按照配管标准规格制造的)　pipe	管子(不按配管标准规格制造的其他用管)　tube
钢管　steel pipe	铸铁管　cast iron pipe
衬里管　lined pipe	复合管　clad pipe
合金钢管　alloy steel pipe	碳钢管　carbon steel pipe
不锈钢　stainless steel pipe	奥氏体不锈钢管　austenitic stainless steel pipe
铁合金钢管　ferritic alloy steel pipe	轧制钢管　wrought-steel pipe
锻铁管　wrought-iron pipe	无缝钢管　seamless (SMLS) steel pipe
焊接钢管　welded steel pipe	电阻焊钢管　electric-resistance welded steel pipe
电熔(弧)焊钢板卷管　electric-fusion (arc)-welded steel-plate pipe	螺旋焊接钢管　spiral welded steel pipe
镀锌钢管　galvanized steel pipe	热轧无缝钢管　hot-rolling seamless pipe
冷拔无缝钢管　cold-drawing seamless pipe	水煤气钢管　water-gas steel pipe
塑料管　plastic pipe	玻璃管　glass tube

续表

橡胶管	rubber tube	直管	run pipe; straight pipe
管道材料规定	piping material specification	管道等级	piping class; piping classification
管道等级号	class designation	公称直径	nominal diameter (DN); nominal (pipe) size
通用连接组	typical installation	双倍加厚的;双倍加强的	double extra heavy; double extra strong
公称压力	nominal pressure (PN)	加厚的;加强的	extra heavy; extra strong
壁厚	wall thickness (WT)	系列号	schedule number (SCH. No.)
管件 Pipe Fitting			
弯头	elbow	异径弯头	reducing elbow
带支座弯头	base elbow	长半径弯头	long radius elbow
短半径弯头	short radius elbow	长半径180°弯头	long radius return
短半径180°弯头	short radius return	带侧向口的弯头(右向或左向)	side outlet elbow (right hand or left hand)
双支管弯头(形)	double branch elbow	三通	tee
异径三通	reducing tee	等径三通	straight tee
带侧向口的三通(右向或左向)	side outlet tee (right hand or left hand)	异径三通(分支口为异径)	reducing tee (reducing on outlet)
异径三通(一个直通口为异径)	reducing tee (reducing on one run)	带支座三通	base tee
异径三通(一个直通口及分支口为异径)	reducing tee (reducing on one run and outlet)	异径三通(两个直通口为异径,双头式)	reducing tee (reducing on both runs, bull head)
45°斜三通	45° lateral	45°斜三通(支管为异径)	45° lateral (reducing on branch)
45°斜三通(一个直通口为异径)	45° lateral (reducing on one run)	45°斜三通(一个口及支管为异径)	45° lateral (reducing on one run and branch)
四通	cross	等径四通	straight cross
异径四通	reducing cross	异径四通(一个分支口为异径)	reducing cross (reducing on one outlet)
异径四通(一个直通口及分支口为异径)	reducing cross (reducing on one run and outlet)	异径四通(两个分支口为异径)	reducing cross (reducing on both outlet)
异径管	reducer	异径四通(一个直通口及两个分支口为异径)	reducing cross (reducing on one run and outlet)
管接头	coupling, full coupling	异径管接头	reducing coupling
半管接头	half coupling	内外螺纹缩接(俗称补芯)	bushing
活接头	union	堵头	plug

续表

管帽 cap	异径短节 reducing nipple; swage nipple
短节 nipple	焊接支管台 weldolet
承插支管台 sockolet	弯头支管台 elbolet
斜接支管台 latrolet	镶入式支管嘴 sweepolet
短管支管台 nipolet	
弯管 Bend	
预制弯管 fabricated pipe bend	跨越弯管(∧形) cross-over bend
偏置弯管(～形) offset bend	90°弯管 quarter bend
环形弯管 circle bend	单侧偏置90°弯管 single offset quarter bend
S形弯管 "S" bend	单侧偏置U形膨胀弯管 single offset "U" bend
U形弯管 "U" bend	双偏置U膨胀弯管 double offset expansion "U" bend
斜接弯管 mitre bend	三节斜接弯管 3-piece mitre bend
偏心异径管 eccentric reducer	同心异径管 concentric reducer
螺纹支管台 threadolet	锻制异径管 reducing swage
折皱弯管 corrugated bend	圆度 roundness
管道特殊件(组件) Piping Specialty (assembly)	
粗滤器 strainer	y型粗滤器 y-type strainer
过滤器 filter	T型粗滤器 T-type strainer
临时粗滤器(锥型) temporary strainer (cone type)	永久过滤器 permanent filter
洗眼器及淋浴器 eye washer and shower	丝网粗滤器 gauze strainer
阻火器 flame arrester	视镜 sight glass
喷嘴;喷头 spray nozzle	取样冷却器 sample cooler
消声器 silencer	膨胀节 expansion joint
波纹膨胀节 bellow expansion joint	单波管 single bellow wave
多波管 multiple bellow	双波管 double bellow
带铰链膨胀节 hinged expansion joint	压力平衡式膨胀节 pressure balanced expansion
自均衡膨胀节(外加强环) self-equalizing expansion joint	轴向位移型膨胀节 axial movement type expansion joint
万向型膨胀节 universal type expansion joint	带接杆膨胀节 tied expansion joint
填函式补偿器 slip type (packed type) expansion joint	球形补偿器 ball type expansion joint

续表

单向滑动填料函补偿器 single actionpacked slip joint	
管道特殊元件、管道支吊架 Piping Special Element, Supports and Hangers	
软管接头 hose connection (HC)	快速接头 quick coupling
金属软管 metal hose	橡胶软管 rubber hose
挠性管 flexible tube	鞍形补强板 reinforcing saddles
补强板 reinforcement pad	特殊法兰 special flange
漏斗 funnel	排液环 drip ring
插板 blank	排液漏斗 drain funnel
8字盲板 spectacle blind; figure 8 blind	垫环 spacer
限流孔板 restriction orifice	爆破片 rupture disk
法兰盖贴面 protective disc	费托立克接头 victaulic coupling
夹钳 clamp	管托 shoe
支耳;吊耳 lug; ear	U形夹(卡) clevis
耳轴 trunnion	锻制U形夹 forged steel clevis
托座 stool	止动挡块 shear lug
带状卡 strap clamp	托架 cradle
夹板 cleat	可调夹板 adjustable cleat
支承环 ring	筋;肋 rib
底板 base plate	加强筋 stiffener
翅片式导向板 finned guide plate	顶板 top plate
管部附着件 pipe attachment	预埋件 embedded part; inserted plate
鞍座 saddle	裙座 skirt
木块 wood block	滑轮组 tackle-block
间隔管(片、块) spacer	支撑杆 strut
杠杆 lever	定位块 preset pieces
带环头拉杆 eye rod	连接杆 connecting rod
垫板(安装垫平用) shim	夹子 clip
连接板 tie plate	连接杆 tie rod
限制杆 limit rod	软管卷盘(筒) hose reel
端部连接 End Connection	
法兰端 flanged end	坡口端 beveled end (BE)

续表

对焊端　butt welded end	平端　plain end (PE)
承插焊端　socket welding end	螺纹端　threaded end (TE)
承口　bell end	焊接端　welding end
法兰连接(接头)　flanged joint	对焊连接(接头)　butt welded joint
螺纹连接;管螺纹连接　threaded joint; pipe threaded joint	万向接头　universal joint
软钎焊连接(接头)　soldered joint	环垫接头　ring joint (RJ)
承插连接(接头)　bell and spigot joint	承插焊连接(接头)　socket welded joint
搭接接头;松套连接　lapped joint	外侧厚度切斜角　bevel for outside thickness
法兰式的　flanged (FLGD)	对焊的　butt welded (BW)
螺纹的　threaded (THD)	承插焊的　socket welded (SW)
小端为平的　small end plain (SEP)	大端为平的　large end plain (LEP)
两端平　both ends plain (BEP)	小端带螺纹　small end thread (SET)
大端带螺纹　large end thread (LET)	两端带螺纹　both end thread (BET)
一端带螺纹　one end thread (OET)	支管连接　branch connection
焊接支管　branch pipe welded directly to the run pipe	
管道用紧固件　Piping Fastener	
紧固件　Fastener	螺栓　bolt
六角头螺栓　hexagonal head bolt	方头螺栓　square head bolt
螺柱;双头螺栓　stud bolt	环头螺栓　eye bolt
沉头螺栓　countersunk (head) screw	地脚螺栓　anchor bolt; foundation bolt
松紧螺旋扣　turnbuckle	U形螺栓　U-bolt
T形螺栓　T-bolt	圆头螺钉　round head bolt
机螺栓;机螺钉　machine bolt	顶开螺栓;顶起螺栓　jack screw
自攻螺钉　self tapping screw	膨胀螺栓　expansion bolt
粗制的　coarse	精制的　refined, fine
螺母　nut	六角螺母　hexagonal nut
方螺母　square nut	蝶形螺母　wing nut
扁螺母　flat nut	锁紧螺母　lock nut
垫圈　washer	平垫圈　plain washer
球面垫圈　spherical washer	弹簧垫圈　spring washer
方垫圈　square washer	斜垫圈　slant washer

续表

销轴	axis pin	销	pin
开口销	cotter pin	定位销	dowel pin
带孔销	pin with hole	圆锥销	taper pin
铆钉	rivet	键	key
图名 Chart Title			
管道布置平面	piping arrangement plan (PAP)	管道布置	piping layout
轴测图	isometric drawing	分区索引图	key plan
初版设备布置图("A"版)	preliminary plot plan ("A" issue)	内部审查版设备布置图("B"版)	internal approval plot plan ("B" issue)
用户审查版设备布置图("C"版)	owner approval plot plan ("C" issue)	确认版设备布置图("D"版)	confirm plot plan ("D" issue)
研究版设备布置图("e"版)	planning plot plan ("E" issue)	设计版设备布置图("F"版)	designing plot plan ("F" issue)
施工版设备布置图("G"版)	construction plot plan ("G" issue)		
厂房、站、单元 Factory, Station and Unit			
分析室	analysis room	蓄电池室	battery room
变压器室	transformer room	控制室	control room
配电室;变电所	substation	通风室	ventilating room
贮藏室	storage room	维修间	maintenance room
办公室	office	压缩机房	compressor house (room)
洗眼站	eye-wash station	泵房	pump house (room)
软管站;公用工程站	hose station (HS); utility station	成套设备	package unit
设施	facilities	泡沫站	foam station
罐区	tank yard	空分装置	air separation facility
工具 Tools			
扭力扳手	torque wrench	液压扳手	hydraulic wrench
液压千斤顶	hydraulic jack	套筒扳手	socket spanner
脚手架	scaffold	厂房顶部的吊车	crane
内径千分尺	inside micrometer	游标卡尺	vernier caliper; sliding caliper
外径千分尺	outside micrometer	表架	dial gauge support/magnetic stand
刻度表	dial gauge	水平仪	level gauge
激光对中仪	laser alignment instrument	万用表	multimeter

续表

塞尺　filler gauge	二硫化钼　molybdic sulfide
管钳子　pip wrench	活扳子　adjustable spanner
弯管气（机）　pipe bender	手锤　hand hammer
弓锯　hack saw	锯子　saw
梯子　ladder	螺丝刀　screw driver
钳套　pincers casing	乐泰　loctite⑦
终端套管（线鼻子）　end sleeve	磨利可　molykote⑧
海洛马　hylomar⑦	锁紧螺母　lock nut
维德加　rivolta⑧	钩环　shackle
内六角　allen key	梅花扳手　ring spanner
冲击扳手　impact wrench	砂轮机　grinding machine
火焰切割机　fire cutting machine	油石　oil stone⑨
灭火毯　fire blanket	丝锥　thread pit/tap⑩
电极夹/焊钳　electrode holder	倍增器　multiplier⑪
乙炔　acetylene	吊架　hanger
热缩管　shrinking hose	弹簧架　spring support
恒力吊架　constant hanger	减振器　snubber
缓冲筒（器）　dash pot	液压减振器　hydraulic snubber
减振装置　damping device	二维限位架　two-axis stop
滚动支架　rolling support	固定架　anchor
限位器　stopper	
单位　Unit	
单位制　system of units	米　meter（m）
毫米　millimeter（mm）	英尺　foot（ft）
英寸　inch（in）	弧度　radian（rad）
度　degree（°）	摄氏　celsius（C）
华氏　fahrenheit（F）	磅/平方英寸　pounds per square inch（psi）
百万帕斯卡　million pascal（MPa）	巴　bar
千克（公斤）　kilogram（kg）	克　gram（g）
牛顿　newton（N）	吨　ton（t）
千磅　kilopound（kip）	平方米　square meter（m²）

续表

方毫米 square millimeter (mm²)	升 liter;litre (L)
立方米 cubic meter (m³)	转/分 revolutions per minute (rpm)
百万分之一 part per million (ppm)	焦(耳) joule (J)
千瓦 kilowatt (kW)	伏(特) volt (V)
安(培) ampere (A)	欧(姆) ohm (Ω)
(小)时 hour (h)	分 minute (min)
秒 second (s)	
应力计算 Stress Calculation	
热应力分析 thermal stress analysis	管道柔性分析 piping flexibility analysis
荷载工况 load case	力 force
力矩 moment of force	反力 reaction
弯曲力矩 bending moment of force	扭矩 torque
荷载 load	外载 externally applied load
工作荷载 working load	冷态荷载 cold load
内力 internal force	外力 external force
力偶 couple of force	管系 piping system
角位移 angular rotation	冷拉 cold spring
位移 displacement	自拉 self spring
节点 node	附加位移 appendant displacement; externally imposed displacement
元件 element	节点号 node number
固定点 fix point; anchor point	安全系数 safety factor
推力 thrust	挠度;弯度 deflection
阻尼振动 damped vibration	耐震等级 seismic class
共振 resonance	机械振动 mechanical vibration
激振;激发 excitation	脉动 pulsation
波谷 wave trough	周期 period
水锤 water hammer	衰减系数 decay factor; decay coefficient; attenuation constant
标准及通用型支架 Standard and Typical Support	
标准管架 standard pipe support	通用管架 typical pipe support
悬臂架 cantilever support	Ⅱ形管架 Ⅱ-type support
跨度 span	组装;装配 assembly

Common Expressions and Vocabularies of Typical Imported Equipment for Energy Engineering 能源工程典型进口设备常用表达及词汇

续表

切割使适合	cut to suit	对中心；找正	alignment
修饰使适合	trim to suit	攻螺孔	tapped；tapping
工程设计阶段及管理 Engineering planning Phase and Management			
分析设计阶段	analytical engineering phase	初步阶段	preliminary stage
成品设计阶段	production design phase	规划布置阶段	planning stage（phase）
基础设计	basic design	详细设计	detail design
开工会议	kick-off meeting；launching meeting	工作程序	working procedure
项目进展情况报告	project status report (PSR)	界区条件	battery limit condition
用户变更通知	client change notice (CCN)	项目审核会	project review meeting
先期确认	advanced certified final (ACF)	厂商协调会	vendor coordinative meeting (VCM)
设计基础数据	basic engineering design data	最终确认	certified final (CF)
工程规定	engineering specification	数据表	data sheet
资料；文件	document	设计文件	design document
附件	appendix	设计规定汇总表	design specification summary sheet (DSSS)
索引；目录	index	技术说明	technical specification
计算表	calculation sheet	工程手册；设计手册	engineering manual
批准用于施工	approved for construction (AFC)	批准用于规划布置	approved for planning (AFP)
详细设计版	detail design issue (DDI)	批准用于设计	approved for design (AFD)
标准	standard (STD)	参考；基准	reference
图	figure；drawing	标准图	standard drawing
参考图	reference drawing	草图	sketch
竣工图	as built drawing	工程图	engineering drawing
待定	hold	说明	description
校核	check	版次	issue
项目	project	项目经理	project manager
状态报告	status report	设计经理	design manager
建设	construction	大修	over haul
模型	model		
化工通用词汇 Chemical Common Words			

续表

absorbent 吸收剂	dimension 尺寸,尺度,维(数),度(数),元
absorber 吸附器	discharge 卸货,流出,放电
accessory 附件	drain 排水沟,消耗,排水
acetylene 乙炔	drawing 图画,制图
action 作用	drum 鼓膜,鼓室
actuator 促动器	dryer 干燥剂
adiabatically 在绝热条件下	drying 烘干
adjuster 调整器	expansion joint 膨胀节
adsorbate 被吸附物	expansion 扩充,开展,膨胀,扩张物
air compressor 空压机	first/second stage condenser 第一/二段冷凝器
balanced opposed compressor 相对平衡式压缩机	first/second stage evaporation 第一/二段蒸发器
blower 风机	first/second/third stage evaporation separator 第一/二/三段蒸发器分离器
bracket 托架	first/second/third stage intercooler 第一/二/三段间冷却器
breaker 断路器	fitting 装配,装置
cap 帽子,(瓶)帽,管帽	floor 地面,地板,基底,(室内的)场地,层
carbon [化]碳(元素符号 C)	flush steam condenser cooler 闪蒸蒸汽冷却器
casting 铸件,铸造	foundation 基础,根本,建立,创立,地基
centrifugal compressor 离心式压缩机	gage(=gauge) 标准度量,计量器
chlorine [化]氯	H_2 converter reactor 脱氢反应器
column 圆柱,柱状物	heat exchanger 热交换器
compressor 压缩机	horizontal 地平线的,水平的
condenser 冷凝器、电容器	HP pressure stripper 高压气提塔
Conditioned cooling water cooler 调温水冷却器	hydrolyzer heat exchanger 水解换热器
conditioned cooling water pump 调温水泵	hydrogen 氢
coupling 联结,接合,耦合,联轴节	hydrolyzer feed pump 水解给料泵
degree 度,程度,[数]次数	hydrolyze 水解
desorber feed pump 解吸给料泵	instrument air compressor 仪表空压机
desorber heat exchanger 解吸换热器	instrument 工具,手段,器械,器具
desorption 解吸,去吸附	integral speed-up gear type isothermal centrifugal compressor 整体增速齿轮式等温离心压缩机
diameter 直径	

续表

joint 接缝,接合处,接合点,连接,接合	reflux condenser 回流冷凝器
level tank for reflux condenser 回流冷凝器液位槽	reflux pump 回流泵
liquid air absorber 液空吸附器	reflux 逆流,回流
liquid oxygen absorber 液氧吸附器	relief 减轻,(债务等的)免除
maximum 最大量,最大限度,极大	restriction 限制,约束
melt 使融化,(使)熔化,使软化	steam cond pump 蒸汽冷凝液泵
minimum 最小值,最小化	steam drum 汽包
MP/LP decomposer 中/低压分解器	steam saturation drum 蒸汽饱和器
MP/LP dissociation heater 中/低压分解加热器	steel 钢,钢铁
multi-stage compressor 多级压缩机	stripper 汽提塔
nickel [化]镍,镍币	turbine cold pump 透平冷凝液泵
nitrogen [化]氮	turbine condenser 透平冷凝器
orifice 孔,口	turbo compressor 透平压缩机
oxygen [化]氧	vacuum 真空,空间,真空吸尘器
plastic 塑胶,可塑体,塑料制品,整形	vertical 垂直的,直立的,顶点的
plug 塞子,插头,插销	volume 体积,量,大量,音量
process condensate cooler 工艺冷凝液冷却器	waster water sump 废水槽
radius 半径,范围,辐射光线,有效航程,界限	waste water sump pump 废水槽泵

① 联轴器用来连接汽轮机、空压机和增压机,带动三个机器一起运转。两端有轮毂,即靠背轮(coupling),中间横着的那一截叫 spacer。但一般都统一叫联轴器 coupling。
② 美国私人公司,提供流体系统元件的销售和服务。一般指管接头、管件等。
③ 阀门部件,一个开位(open limit switch),一个关位(close limit switch)。
④ 也称失气开/失气关,指阀门在气源故障时打开/关闭。
⑤ 信号选择器,高选器选择最高信号,低选器选择最低信号。
⑥ 常架在汽轮机两端轴承箱,检查轴承箱是否跑偏。
⑦ (音译)都是密封胶,但用的地方不一样。
⑧ 都是润滑脂,如用在蒸汽入口法兰的螺母内。
⑨ 打磨金属表面。
⑩ 用于修复受损螺纹。
⑪ 力矩转换。

References
参 考 文 献

［1］ 上海市职业技术教育课程改革与教材建设委员会. 机电与数控专业英语［M］. 北京：机械工业出版社，2002.
［2］ Amanda Crandell Ju. 商务英语情景口语 100 主题［M］. 北京：外文出版社，2007.
［3］ 李光布. 机械工程专业英语［M］. 武汉：华中科技大学出版社，2008.
［4］ 段荣娟. 矿业实用英语［M］. 徐州：中国矿业大学出版社，2009.
［5］ 杨庆彬，谢安全，王胜春. 英汉·汉英煤化工词汇［Z］. 北京：化学工业出版社，2011.
［6］ 陈国桓，蔡晖. 英汉·汉英化工工艺与设备图解词典［Z］. 第 2 版. 北京：化学工业出版社，2017.